Small Supernumerary Marker Chromosomes (sSMC)

Thomas Liehr

Small Supernumerary Marker Chromosomes (sSMC)

A Guide for Human Geneticists and Clinicians

With contributions by Unique
(The Rare Chromosome Disorder Support Group)

 Springer

Dr. Thomas Liehr
Universitätsklinikum Jena
Institut für Humangenetik
Kollegiengasse 10
07743 Jena, Germany
i8lith@mti.uni-jena.de

UNIQUE
(The Rare Chromosome Disorder Support Group)
P.O. Box 2189
Caterham, Surrey
CR3 5GN, UK
info@rarechromo.org
www.rarechromo.org

ISBN 978-3-642-43536-2   ISBN 978-3-642-20766-2 (eBook)
DOI 10.1007/978-3-642-20766-2
Springer Heidelberg Dordrecht London New York

Author's Disclaimer: The clinical details given for specific chromosomal imbalances, including such regions causing, according to present knowledge, no harm, represent the presently available data. They can be used for interpretation of cytogenetic findings – however, there are always exceptions from the findings to be expected. Some are described in this book. Thus, please use this information carefully! The author does not take any responsibility for (mis)interpretation of the data provided in this book.

Cover design: deblik, Berlin

Printed on acid-free paper

Springer is part of Springer Science+Business Media (www.springer.com)

Foreword

When Unique started up in 1984 as the Trisomy 9p Support Group, there was virtually no information or support for families about any rare chromosome disorder. Today it is different for people diagnosed with a known syndrome, but for the majority, including most people with a small supernumerary marker chromosome, little has changed. With its ever-increasing membership – currently standing at more than 10,000 individuals in 80 different countries – Unique fills that gap.

Small Supernumerary Marker Chromosomes is a welcome collaboration between a leading scientist and a family support group to create an up-to-date picture of one type of rare chromosome disorder. Scientific and clinical reports are brought to life by families' descriptions of the consequences of having a child with a small extra chromosome. Eighteen Unique families tell you in words and photographs what having this rare chromosome disorder means. Most of the children's names have been changed in accordance with their parents' wishes.

Unique
The Rare Chromosome Disorder Support Group
PO Box 2189
Caterham, Surrey
CR3 5GN, UK
http://www.rarechromo.org

Preface

Since 1992 I have been working in the field of clinical cytogenetics. My diploma, i.e., a master's thesis, was about a special subgroup of patients with small supernumerary marker chromosomes (sSMC), the cat eye syndrome (Liehr et al. 1992). Since that time much progress has been achieved in the field of sSMC. Especially the sSMC homepage (Liehr 2011) with presently more than 4,000 single sSMC case reports together with the advance of technical possibilities for a comprehensive characterization of this special group of rearranged chromosomes enables today much better genotype–phenotype correlations than when I started to study sSMC.

Nonetheless, I recently met a family with the following story, providing evidence that lots of knowledge on sSMC that is nowadays available did not reach the public health system as it should. An sSMC was detected after amniocentesis in the fetus of a pregnant woman who was referred for cytogenetic analysis because of advanced maternal age; sonographic findings were normal. The gynecologist told the couple that the cytogenetic finding was connected with an adverse prognosis and that the developing child would be "100% disabled and mentally retarded." The parents thus terminated the pregnancy. Later, it turned out that the sSMC was not only parentally derived but also that the first healthy child of the couple also had the same sSMC. This book is intended to help avoid similar situations and to be informative to clinicians, cytogeneticists, and families.

Besides the present knowledge on sSMC, including the biological background, also clinically relevant information is included together with personal reports of families having a child affected with an sSMC. The latter was realized in close collaboration with Unique, the Rare Chromosome Disorder Support Group (http://www.rarechromo.org/), and by contributions provided by families in contact with the author.

Jena, October 2011   Thomas Liehr

References

Liehr T (2011) The sSMC homepage. http://www.med.uni-jena.de/fish/sSMC/00START.htm. Accessed 10 Jan 2011, Also accessible via http://markerchromosomes.wg.am or http:// markerchromosomes.ag.vu

Liehr T, Pfeiffer RA, Trautmann U (1992) Typical and partial cat eye syndrome: identification of the marker chromosome by FISH. Clin Genet 42:91–96

Acknowledgments

This book would not have been possible without the support of the families telling their stories. Furthermore, the sSMC research of the author was supported during recent years by the following foundations: Deutsche Forschungsgemeinschaft (DFG; project numbers 436 RUS 17/109/04, 436 WER 17/5/05, LI 820/22-1, and LI 820/332-1), Else-Kröner-Fresenius-Stiftung (2011_A42), Deutscher Akademischer Austauschdienst (DAAD, project numbers 313-ARC-XX-lk, 324-04jo, and A0703172/Ref.325), Dr. Robert Pfleger Stiftung, Scheringstiftung, Herbert Quandt Stiftung der VARTA AG, Evangelisches Studienwerk e.V. Villigst, Böhringer Ingelheim Fonds, and Erwin Riesch-Stiftung.

Contents

Abbreviations

AC	Accessory chromosome
aCGH	Array-based comparative genomic hybridization
ACH	Accessory chromosome
AS	Angelman syndrome
BWS	Beckwith–Wiedemann syndrome
caNC	Cancer-associated neochromosome
CBG	C bands by barium oxide using Giemsa stain
CES	Cat eye syndrome
CGH	Comparative genomic hybridization
CVS	Chorionic villus sampling
der	Derivative chromosome
ES	Emanuel syndrome
ESAC	Extra structurally abnormal chromosome
FISH	Fluorescence in situ hybridization
GTG	G bands by trypsin using Giemsa stain
hUPD	Heterodisomy
i18pS	Isochromosome 18p syndrome
inv dup	Inverted-duplication-shaped small supernumerary marker chromosome
ISCN	International System for Human Cytogenetic Nomenclature
iUPD	Isodisomy
LCR	Low copy repeat
Mb	Megabase
min	Centric minute-shaped small supernumerary marker chromosome
NMC	Neocentric marker chromosome
NOR	Nucleolus organizing region
OMIM	Online Mendelian Inheritance in Man
p	Short chromosome arm
PCR	Polymerase chain reaction
PKS	Pallister–Killian syndrome
PWS	Prader–Willi syndrome

q	Long chromosome arm
r	Ring chromosome
SAC	Small accessory chromosome
SBAC	Small bisatellited additional chromosome
SMRC	Supernumerary minute ring chromosome
SMC	Supernumerary marker chromosome(s)
SRC	Supernumerary ring chromosome
SRS	Silver–Russell syndrome
sSMC	Small supernumerary marker chromosome(s)
TND	Transient neonatal diabetes
TS	Turner syndrome
UBCA	Unbalanced chromosomal abnormality
UPD	Uniparental disomy

Chapter 1
Introduction

The topic of this book is small supernumerary marker chromosomes (sSMC). Below, sSMC will be introduced, defined, and information will be given about their nomenclature, shapes, the frequencies with which they appear, and their effects. But first we have to understand the field we are entering by focusing on sSMC. So, in general we are talking here about a group of patients identifiable by human genetic diagnostics, especially by cytogenetics. The latter is a branch of genetics concerned with the study of the structure and function of chromosomes.

The cytogenetic era started when the first chromosomes were visualized through microscopes, which was around 1880 (Arnold 1879). However, the first really evaluable human metaphase chromosome spread for diagnostics was published almost six decades ago (Tjio and Levan 1956). Interestingly, the determination of the correct modal human chromosome number (Tjio and Levan 1956), the detection of the first chromosomal abnormalities such as Down syndrome (Lejeune 1959), and the detection of the presence of sSMC (Ilberry et al. 1961) occurred before the invention of the Q-banding method by Dr. Lore Zech from Uppsala, Sweden (Caspersson et al. 1968). The ability to generate a black-and-white banding along the chromosomes enabled the detection of more abnormalities, such as translocations, inversions, deletions, and insertions. Nowadays, the G bands by trypsin using Giemsa stain (GTG) banding approach (Seabright 1971) is still the starting point and gold standard of all cytogenetic techniques (see Sect. 4.1). It is relatively cheap, easy to perform, and gives an overview of the whole human genome; the resolution is limited to some ten million base pairs (Mb).

The pure chromosome banding era ended in 1986 with the first so-called "molecular cytogenetic" experiment on human chromosomes, which was at the same time the starting point of the youngest discipline in human genetic diagnostics (for a review see Liehr and Claussen 2002; see Sect. 4.1). The major technique in molecular cytogenetics is fluorescence in situ hybridization (FISH; for reviews see Liehr and Claussen 2002; Liehr 2009a). FISH is an approach that allows nucleic acid sequences to be examined inside cells or on chromosomes, and was described first in 1986 for humans (Pinkel et al. 1986). Between 1986 and 1996 one-color to three-color FISH experiments were performed, but since 1996 multicolor FISH probe sets have become more and more important in routine cytogenetics (for reviews see Liehr et al. 2006b; Liehr 2011b). Recently, the so-called array techniques (Forster et al. 2003) were introduced in cytogenetic diagnostics, providing high resolution for the determination of chromosomal breakpoints. Particularly, array-based comparative genomic hybridization (aCGH; for reviews see Tabor and Cho 2007; Liehr 2009a), which is a refined molecular cytogenetic technique (chromosome-based comparative genomic hybridization, CGH; Kallioniemi et al. 1992), is widely used nowadays for sSMC characterization (Backx et al. 2007; Pietrzak et al. 2007; Tsuchiya et al. 2008).

Overall, cytogenetics means chromosomal analysis at the single-cell level. In contrast, molecular analysis of DNA, including aCGH, and other new whole-genome-directed approaches such as "next- or second-generation sequencing" (ten Bosch and Grody 2008) are investigations of millions of cells. Thus, cytogenetics and molecular genetics can complement and assist each other to achieve the goal of characterizing aberrant karyotypes as comprehensively as possible. For sSMC detection and characterization, cytogenetic analysis is in almost all cases the initial step.

1.1   The Problem

Marker chromosomes, in general, are according to the International System for Human Cytogenetic Nomenclature (ISCN) "structurally abnormal chromosomes of unknown origin, frequently found in karyotypes of cancer patients and patients with constitutional genetic disorders" (ISCN 2009). This definition of markers includes a conglomerate of differently shaped and sized chromosomes, including those this book is about. However, sSMC are something special, as outlined below.

In older literature, the report most often cited as the first description of an sSMC is that of Froland et al. (1963), describing an isochromosome 18p (Rivera et al. 1984). However, this was in fact the third sSMC report, preceded by the report of Ellis et al. (1962), who reported "an aberrant small acrocentric chromosome," and that of Ilberry et al. (1961), who reported the first sSMC ever.

1.1.1   Definition of sSMC

Up to 2003 a clear definition of sSMC was lacking throughout the literature. A "minimal definition" for sSMC was that they are "small structurally abnormal chromosomes that occur in addition to the normal 46 chromosomes" (Crolla et al. 1997). Part of the problem with definition is that the phenotypes associated with sSMC are hugely variable, from normal to severely abnormal (Paoloni-Giacobino et al. 1998). Additionally, sSMC are a morphologically heterogeneous group of structurally abnormal chromosomes (see Sect. 1.1.3). Thus, the denomination of sSMC as "markers" makes sense and should be maintained even after their identi-fication by molecular cytogenetics, even though ISCN (2009) recommends the use of "derivative" if the chromosomal origin is known.

Nowadays, sSMC can be defined cytogenetically as *structurally abnormal chromosomes that cannot be identified or characterized unambiguously by conventional banding cytogenetics alone, and are (in general) equal in size to or smaller than a chromosome 20 of the same metaphase spread.* sSMC can be present additionally (1) in a karyotype of 46 normal chromosomes, (2) in a numerically abnormal karyotype (e.g., Turner syndrome or Down syndrome), or (3) in a structurally abnormal but balanced karyotype (e.g., Robertsonian translocation; Wolff and Schwartz 1992) or ring chromosome formation (Baldwin et al. 2008).

In contrast, an sSMC larger than chromosome 20 can usually be identified on the basis of chromosome banding. Thus, the definition of sSMC vs. large(r) supernumerary marker chromosomes (SMC) is a cytogenetic not a "functional" one, i.e., sSMC and larger SMC can have the same modes of formation and karyotypic evolution (Liehr et al. 2004; see Chap. 3).

1.1.2 Nomenclature

sSMC have been given various names over the last few decades. The three best known are "SMC," which does not distinguish between larger and smaller SMCs, extra structurally abnormal chromosome (ESAC), and supernumerary ring chromosome (sSRC). In addition, the following names can be found: accessory chromosome (AC or ACH), small accessory chromosome (SAC), marker chromosome (MC), extra or additional marker chromosome, supernumerary or extra microchromosome, additional or metacentric chromosome fragment, (centric) fragment, small bisatellited additional chromosome (SBAC), neocentric marker chromosome (NMC), supernumerary minute ring chromosome (SMRC), and cancer-associated neochromosome (caNC) in exceptional cases where an sSMC is acquired in neoplasia (reviewed in Liehr et al. 2004; Liehr 2011a).

This flood of names is presently tending to be replaced by the use of sSMC, facilitating tremendously a literature search.

1.1.3 Shapes of sSMC

sSMC can have three different shapes (Fig. 1.1). There are (1) inverted duplicated (shaped) sSMC, abbreviated in the karyotype formula as "inv dup" (2) ring-shaped sSMC, written as "r" in the karyotype, and (3) centric minute-shaped sSMC. The latter are most often referred to as "min" or also as "del" (for deleted in parts) or as "der" (for derivative). As outlined in Liehr (2009b) there is ongoing discussion about the correct karyotypic nomenclature for sSMC.

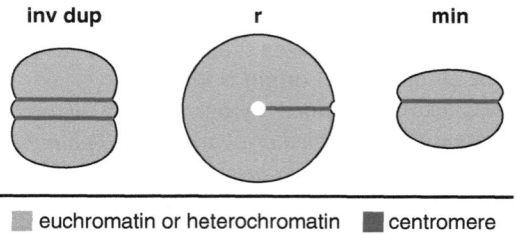

Fig. 1.1 Small supernumerary marker chromosomes (sSMC) can have three basic shapes: inverted duplication (*inv dup*), ring (*r*), and centric minute (*min*). They can be constituted by euchromatin and/or heterochromatin and have a primary constriction, i.e., a centromere. If the shape of an sSMC was not studied in detail, it can be described in a karyotype as "mar" for marker chromosome

The three basic shapes (Fig. 1.1) are present in all sSMC subgroups: centric sSMC including alphoid DNA as a centromere-forming region (Liehr et al. 2006b; see Chaps. 5 and 6), neocentric (Liehr et al. 2007a; see Chap. 7), multiple (Liehr et al. 2006c; see Chap. 8) and complex rearranged (Trifonov et al. 2008; see Chap. 10) sSMC, sSMC present in individuals with TS karyotypes (Liehr et al. 2007b; see Sect. 5.5), and sSMC in individuals with other imbalances such as trisomy 21 and imbalances of X or of other chromosomes (Liehr 2011a; see Chap. 9). According to the sSMC homepage (Liehr 2011a) the most frequently observed sSMC in terms of shape are inverted duplicated sSMC (63%), followed by centric minute-shaped sSMC (26%) and ring-shaped sSMC (11%).

1.2 Frequency of sSMC

The sSMC frequencies of four different human subpopulations are known: (1) the prenatal population, (2) the postnatal population, (3) patients with fertility problems, and (4) patients with developmental and/or mental retardation. The data are based on 132 corresponding studies on approximately 1.3×10^{6} individuals, 980 of whom had an sSMC (Liehr and Weise 2007; Table 1.1).

1.2.1 Normal Population

1.2.1.1 Normal Population: Prenatal

sSMC are present in 0.075% of "unselected" prenatal individuals and only in 0.044% of consecutively studied postnatal ones (Liehr and Weise 2007; Table 1.1). The apparently reduced rate might be in part connected with the fact that 30–50% of pregnancies with the diagnosis "child with an sSMC" are terminated

Table 1.1 sSMC frequency in different human subpopulations according to Liehr and Weise (2007)

Human population studied	Number of cases	Cases with sSMC Absolute number	Cases with sSMC Percentage
Prenatal cases (consecutively collected)	688,030	514	0.075
Preselected prenatal			
de novo sSMC	377,357	162	0.043
in ultrasonography aberrant	4,409	9	0.204
intracytoplasmic sperm injection	4,625	2	0.043
Newborn cases (consecutively collected)	121,694	54	0.044
Healthy adults	1,405	1	0.071
Males with fertility problems	21,841	36	0.165
Females with fertility problems	9,165	2	0.022
(Develop)mentally retarded	69,332	200	0.288

(Warburton 1991; Kumar et al. 1997; Cavani et al. 2003; Shaffer et al. 2004). As Cavani et al. (2003) showed, 100% of pregnancies with inherited sSMC (see Chap. 2) were continued, whereas almost 50% of those with de novo sSMC (see Chap. 2) were terminated. Thus, in a certain percentage of cases, pregnancies that would have produced healthy children with (de novo) sSMC are interrupted.

Furthermore, according to Liehr and Weise (2007), it can be concluded that there seems to be no significant: a natural loss of pregnancies in connection with sSMC between the second trimester and birth. However, as the sSMC rate in newborns of 0.044% is only just below half of the prenatally detected one, this highlights that in prenatal diagnostics a preselected subpopulation of humans is being studied. In conclusion, the rate of 0.075% of sSMC carriers in prenatal studies is biased by three main factors: a higher rate of cases with sSMC in prenatal individuals compared with newborns can be due to (1) the maternal age effect in prenatal series (Crolla et al. 2005), (2) the fact that invasive prenatal diagnosis is carried out in up to one third of the cases due to known or suspected fetal pathology, and/or (3) severely affected fetuses may result in miscarriage and will therefore not be included among newborns: 4.4% of sSMC pregnancies end with stillbirth or spontaneous abortion (Kumar et al. 1997). One has to be aware that no data are available on the unbiased rate of sSMC carriers in the general prenatal human population; all that is available is the frequency in the prenatally studied human population.

Not surprisingly, the sSMC rate of 0.204% for the prenatal population with abnormalities detected by ultrasonography is almost five times higher than that for the newborn population and almost three times higher than that for the prenatal population in general (Liehr and Weise 2007; see Table 1.1). Among such cases are the 4.4% of sSMC pregnancies which end in stillbirth or spontaneous abortion (Kumar et al. 1997). Also, sSMC carriers with a clinically abnormal outcome are included in this group. As listed in Table 1.1, 0.288% of (develop)mentally retarded patients have an sSMC. Interestingly, no increased risk of sSMC presence was detected in pregnancies induced by intracytoplasmic sperm injection (Liehr and Weise 2007).

1.2.1.2 Normal Population: Postnatal

As no ethnic effects in sSMC frequency are known (Liehr and Weise 2007), among the approximately seven billion humans there should be approximately 3.1×10^6 sSMC carriers and 2.2×10^6 of them should have a de novo sSMC (Liehr 2011a). Approximately 70% of de novo sSMC carriers (Graf et al. 2006) and over 98% of inherited sSMC carriers are clinically normal (Crolla 1998; for de novo and inherited sSMC see Chap. 2). That is, most clinically normal sSMC carriers (approximately 2.4×10^6) do not and will never know that they have this genetic condition. In these individuals there are two possible scenarios in which an sSMC is diagnosed. It may be found purely by chance when cytogenetic analysis is performed, e.g., as a parental chromosome analysis because of a child with another chromosomal rearrangement, or by tumor cytogenetic analysis in connection with leukemia diagnostics (Liehr 2011a). The other possibility is that an sSMC is detected in couples studied cytogenetically because of fertility problems (Manvelyan et al. 2008; Sect. 1.2.2).

1.2.2 Healthy Population with Fertility Problems

It is well known that 0.125% of people with problems conceiving are sSMC carriers (Chandley et al. 1975; Manvelyan et al. 2008). Distinguishing male from female, there seems to be a 7.5:1 difference in the sSMC frequency for this special group (Liehr and Weise 2007; Table 1.1). However, this might be an ascertainment bias because in another study summarizing 111 sSMC picked up in connection with fertility problems and characterized in detail by molecular cytogenetics, 41% of patients were female (Manvelyan et al. 2008).

In males there are hints that oligozoospermia is significantly correlated with sSMC presence in 7% of cases, whereas in azoospermia sSMC are present in less than 1% of the relevant cases (Mau-Holzmann 2005). According to Manvelyan et al. (2008), in 60% of cases with "infertility" the sSMC originates from chromosome 14 or 15. Also euchromatic imbalances are caused by sSMC presence in 30% of cases. Most strikingly, in 53% of "infertile" sSMC carriers the sSMC was parentally transmitted (see Chap. 2). Manvelyan et al. (2008) showed that fertility problems were more frequent in males inheriting the sSMC from the mother than from the father, and for females the reverse relationship was found.

Overall, sSMC may be just a chance finding in a certain subset of healthy persons with fertility problems. Nevertheless the gender of the sSMC carrier has to be considered. In males, sSMC presence seems to have a more adverse influence on fertility than in females (see Sect. 2.1).

1.2.3 Developmentally and Mentally Retarded Persons

In (develop)mentally retarded patients the sSMC rate is elevated to 0.288%, i.e., sSMC can be found approximately seven times more frequently than in the general population. This is not surprising and is seen in connection with the fact that patients with Emanuel syndrome (ES; see Sect. 5.1), Pallister–Killian syndrome (PKS; see Sect. 5.3), cat eye syndrome (CES; see Sect. 5.2), isochromosome 18p syndrome (i18pS; see Sect. 5.4), and also patients with larger inverted duplicated chromosomes derived from chromosome 15 (see Sect. 6.15) are overrepresented in this clinical group (Liehr and Weise 2007).

1.3 Chromosomal Origin of sSMC

As mentioned in Sect. 1.2, for sSMC frequencies one has to be aware that no data are available on the real sSMC rate of chromosomal origin within humans – only the studied and published cases can be taken as a basis for this kind of estimation. However, it is quite clear that sSMC are most often derived from chromosome 15, followed by chromosome 22 (Fig. 1.2).

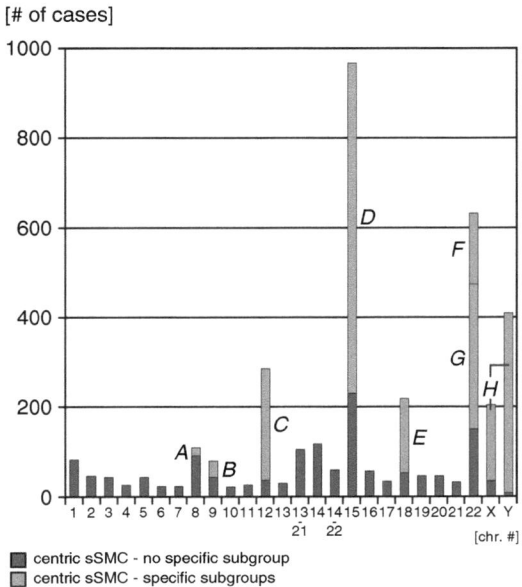

Fig. 1.2 Chromosomal origin of sSMC according to the sSMC homepage (Liehr 2011a). For technical reasons, in some cases it is not possible to distinguish if an sSMC is derived from chromosome 13 or 21 and chromosome 14 or 22; the frequencies of these groups are given under 13–21 and 14–22. # number; *A–H* cases with following syndromes: isochromosome 8p syndrome (*A*), isochromosome 9p syndrome (*B*), Pallister–Killian syndrome (PKS) (*C*), isochromosome 15 syndrome (*D*), isochromosome 18p syndrome (*E*), cat eye syndrome (CES) (*F*), Emanuel syndrome (ES) (*G*), and Turner syndrome (TS) (*H*)

1.3.1   sSMC in Individuals with 47 Chromosomes

Among individuals with karyotype 47,XN,+mar, the sSMC is most often derived from chromosome 15 (approximately 30%), followed by chromosome 22 (approximately 20%). In sSMC with a chromosome 22 origin, ES cases (approximately 10%) and CES cases (approximately 5%) can be found. Overall, approximately 60% of sSMC derive from acrocentric chromosomes within this group. For non-acrocentric-derived sSMC, chromosome 12 (approximately 9%) and chromosome 18 (approximately 7%) are the most frequent origins, as they include PKS patients (approximately 8%) and i18pS patients (approximately 5%). The remainder (34%) of this group of sSMC are distributed across the other human chromosomes. For nonacrocentric sSMC, sSMC are most frequently derived from chromosome 8 (approximately 3%) and chromosome 1 (approximately 2.5%).

1.3.2   sSMC in Individuals with 46 Chromosomes: Turner Syndrome Karyotype Cases

For the special group of patients who have a TS mosaic karyotype of 46,X,+mar, mostly derivatives of X (approximately 69%) and Y (approximately 30%) chromosomes can be observed (Fig. 1.2; see Sect. 5.5). There is also a very rare subgroup of TS patients constituting approximately 1% where the sSMC is derived from any of the autosomes and not from a gonosome.

1.3.3   Multiple sSMC

Patients with multiple sSMC show karyotypes of 48,XN,+mar1,+mar2 up to 53,XN, +mar1,+mar2,+mar3,+mar4,+mar5,+mar6,+mar7. All two to seven sSMC are derived from different chromosomes. In general, in multiple sSMC a different chromosomal distribution is present compared with those sSMC discussed in Sects. 1.3.1 and 1.3.2 (Fig. 1.3). For chromosome 6 an overrepresentation is obvious in multiple sSMC (Liehr et al. 2006c), whereas chromosomes 15 and 22 are underrepresented. This supports the hypothesis that multiple sSMC derived from different chromosomal regions evolve differently to single sSMC during gameto-genesis or embryogenesis (Beverstock et al. 2003; see Sect. 3.6).

1.3.4   Neocentric sSMC

The most frequently observed neocentric sSMC (see Chap. 7) are derived from chromosomes 15, 13, 8, 3, and 1. It was speculated that neocentric sSMC and centric sSMC have in part a common origin (Liehr et al. 2004, 2007a). However, this idea was not confirmed by a recent study by Murmann et al. (2009). No neocentric sSMC derived from chromosomes 19, 21, and 22 have been reported,

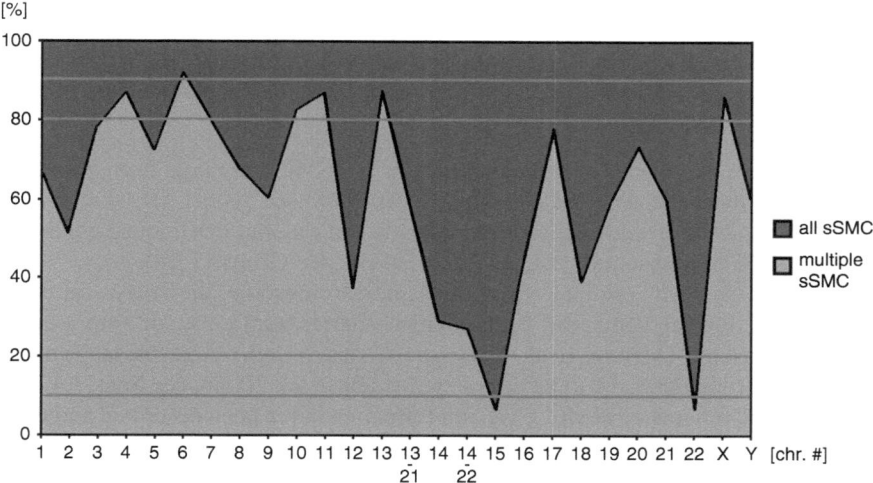

Fig. 1.3 The chromosomal origin of sSMC in general is compared with multiple sSMC descent. It is obvious that chromosome 6 is overrepresented and that chromosomes 15 and 22 are underrepresented in multiple sSMC. Other chromosomes such as chromosomes 4, 7, 10–11, 13, 17, and 20 and possibly the X chromosome could also be overrepresented in multiple sSMC

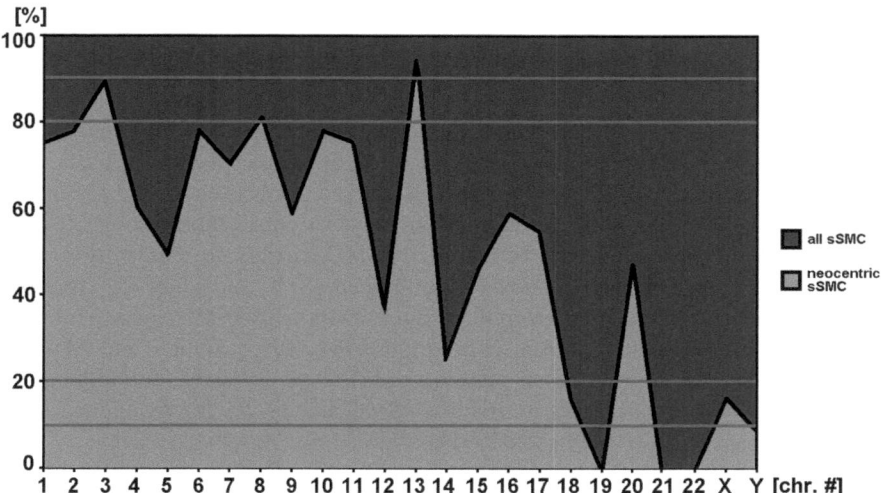

Fig. 1.4 The chromosomal origin of sSMC in general and of neocentric sSMC is compared; neocentric sSMC derived from chromosomes 3 and 13 are overrepresented and those derived from chromosomes 19, 21, and 22 are underrepresented, i.e., have not been reported yet

which may be due to lack of corresponding sites for neocentromere formation on these particular chromosomes (Fig. 1.4) (for reviews on neocentromeres see Warburton 2004; Stanyon et al. 2008).

1.4   What Are the Effects of sSMC

There are several reasons why it is difficult to answer the simple question of what the effects of sSMC are:

1. Phenotypes associated with the presence of an sSMC can range from normal to severely abnormal (Paoloni-Giacobino et al. 1998; see Sect. 1.1)
2. sSMC can be derived from any chromosome and can, but must not be combined with other chromosomal changes (Liehr 2011a; see Chaps. 5–10)
3. Even if two sSMC originate from the same chromosome, they may still differ in size and in the content of euchromatic material from either or both arms of a chromosome (Maurer et al. 2001; see Chaps. 5–10);   Also mosaicism and cryptic mosaicism have to be considered (Liehr et al. 2006b; see Sect. 1.4.3)
4. Structural variants of sSMC, e.g., ring formation, have been described and there might be effects arising from these (von Beust et al. 2005; see Sects. 1.1.3 and 1.4.1)
5. Some patients have multiple sSMC of different origin (Liehr et al. 2006c; see Sect. 1.3.3 and Chap. 8)
6. One cannot be sure that an sSMC is really correlated with and the reason for a clinical abnormality detected in a certain patient (Nelle et al. 2010; see Chap. 9)
7. Silent euchromatin duplications have been described – what about genetic silencing in sSMC? (Stephane and Genevieve 1999; see Sect. 1.4.1)

Irrespective of all these points, the following data are available on the general effects of sSMC presence:

1. In approximately 70% of cases a de novo sSMC has no phenotypic effects (Graf et al. 2006). More specifically, the risk of an abnormal phenotype associated with a de novo sSMC is 7% if the sSMC is derived from chromosome 13, 14, 21, or 22 and 28% if the SMC is coming from nonacrocentric chromosomes (Crolla 1998). Also, more than 98% of inherited sSMC carriers are clinically normal (Baldwin et al. 2008; for de novo and inherited sSMC see Chap. 2). Approximately 16% of prenatally ascertained individuals with sSMC show abnormal phenotypes (summarized from Warburton 1984, 1991; Daniel and Malafiej 2003). The overall risk of an abnormal phenotype is 10.9% for individuals with satellited (i.e. acrocentric derived) sSMC and 14.7% for individuals with nonsatellited sSMC (Warburton 1991).
2. It is known that sSMC may lead to reduced fertility in males without additional clinical sSMC-related symptoms (Chandley et al. 1975; Manvelyan et al. 2008; see Sect. 1.2.2). However, nothing is really known about the mechanism causing this kind of infertility. In 1974 it was postulated that a bisatellited chromosome by association with acrocentric chromosomes may interfere with mitosis at a critical stage of fetal development (Friedrich and Nielsen 1974; Kumar et al. 1997). Chandley et al. (1975) suggested that sSMC could disrupt human sper-matogenesis. Furthermore, there is one study on a male with normal sperm (normozoospermia) and an sSMC in 100% of peripheral blood cells.

Here sSMC presence was proven in only 26% of the sperm and 42% of the fertilized embryos studied (Oracova et al. 2009). Overall, this supports the idea of evolutionary pressure against sSMC within the male germ line (Liehr 2006; Liehr et al. 2008a; see Chap. 2).

3. There are some well-defined sSMC-related syndromes with a specific expected range of symptoms. These are especially PKS (see Sect. 5.3), CES (see Sect. 5.2), ES (see Sect. 5.1), i18pS (see Sect. 5.4), isochromosome 15 syndrome (see Sect. 6.15), and to a certain extent TS (see Sect. 5.5).

4. Finally, for those sSMC where no "typical sSMC-related syndromes" such as PKS, CES, ES, and i18pS have been defined yet, there are three main features that determine the effects of sSMC presence: gain of copy number (see Sect. 1.4.1), uniparental disomy (UPD; see Sect. 1.4.2), and mosaicism (see Sect. 1.4.3).

1.4.1 Gain of Copy Number

In sSMC diagnostics, it is most important to clarify the chromosomal origin of the extra chromosome. The next step in diagnostic procedure is the characterization of the pericentric region the sSMC consists of (see Chap. 4). If no genetically relevant material can be detected, in many instances this means that the sSMC will have no direct clinical impact on the phenotype of its carrier, even though UPD (see Sect. 1.4.2) and mosaicism (see Sect. 1.4.3) have to be considered.

A patient with an sSMC that consists of euchromatin[1] and heterochromatin[2] has to be evaluated in detail for the meaning of the genetic imbalance induced by the extra chromosome. Unbalanced chromosomal abnormalities (UBCA) such as an sSMC do not necessarily lead to clinical phenotypes, as recently reviewed and summarized (Barber 2005, 2011). The fact that even euchromatic imbalances can be correlated with a clinically normal outcome is obvious when sSMC cases are reviewed by chromosome (Mrasek et al. 2005; Liehr et al. 2006b; Liehr 2011a; see Chap. 6). Thus, for each pericentric region there are specifically sized chromosomal regions which can be present in one or more copies without an adverse effect (Fig. 1.5). However, a dose-sensitive region begins at a certain distance distal to the heterochromatic centromere and a UBCA then leads to clinical abnormalities in the affected person. As well as the size of the UBCA, copy number can also be important as chromosomal regions have been identified which can be compensated for if they are present in three copies (e.g., trisomy of the short arm of chromosome 18; Niksic et al. 2010), but which lead to a specific sSMC-related syndrome if present fourfold (i18pS with tetrasomy of the short arm of chromosome 18;

---

[1]Euchromatin is genetic material including known actively transcribed/translated genes.

[2]Heterochromatin is genetic material without (active) genes.

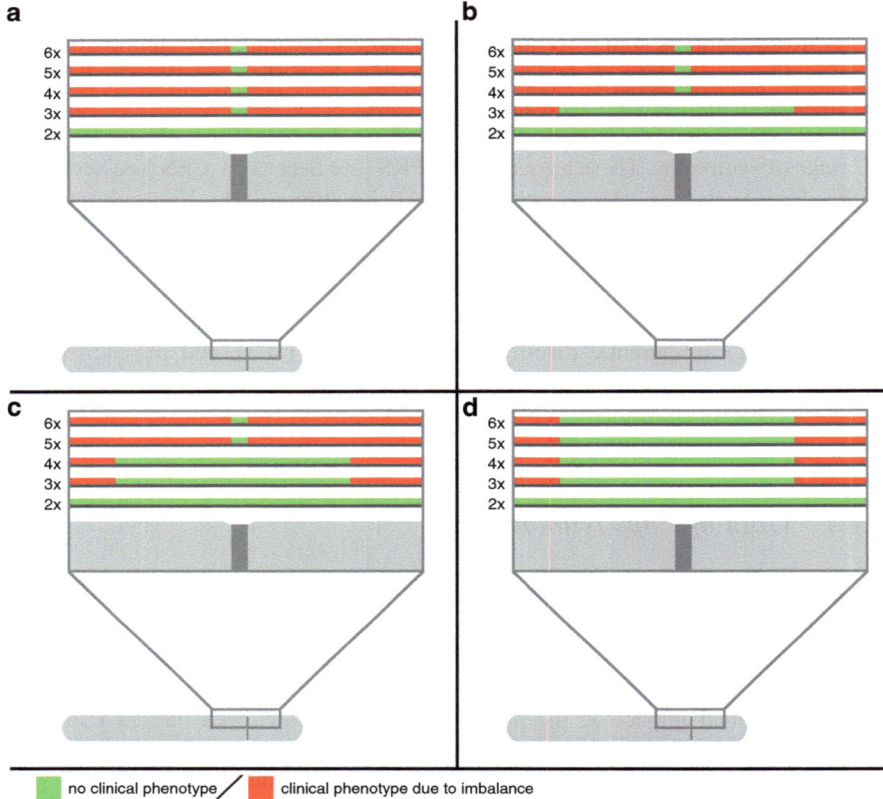

Fig. 1.5 sSMC presence leads to gain of copy number in the pericentric region of one or more chromosomes. The gain can be between threefold and sixfold. Here different clinical phenotypes due to sSMC presence are shown schematically for four scenarios. For all four of them, examples can be found (Liehr 2011a). (**a**) The presence of additional euchromatic material besides the heterochromatic centromeric region leads to clinical problems. (**b**) One additional copy of a centromere near material of a certain size can be tolerated. If the extent of euchromatic material is too large or if the euchromatic material is present between fourfold and sixfold, an abnormal phenotype is the consequence. (**c**) Same situation as in (**b**); however, even four copies of the corresponding region can be compensated for, but not five or six of them. (**d**) Irrespective of copy numbers between threefold and sixfold, a certain size of chromosomal imbalance due to sSMC presence can be tolerated without clinical consequences

Rivera et al. 1984). Copy number gain due to sSMC can range between three- and sixfold (Nietzel et al. 2003; Fig. 1.5).

A phenomenon called genetic silencing might also play a role for the clinical consequences of the presence of sSMC (Stephane and Genevieve 1999). As highlighted before, sSMC consist of euchromatin and heterochromatin (Fig. 1.1). Heterochromatin according to Hahn et al. (2010) is a repressively acting chromosome state that is characterized by densely packed DNA and low transcriptional

activity, and can induce gene silencing in neighboring euchromatin. Alterations in
the proportion of euchromatin to heterochromatin can principally impair normal
gene expression patterns (Hahn et al. 2010). Within an sSMC the proportion of
heterochromatin to euchromatin is shifted much more toward heterochromatin than
in a normal, non-rearranged chromosome. Thus, silencing/heterochromatization
might play a role especially in a situation such as that depicted schematically in
Fig. 1.5d. However, this has not been verified in any studies.

1.4.2   Uniparental Disomy

UPD is the presence of a chromosome pair derived only from one parent (maternally or paternally derived) within a disomic cell line. The UPD concept itself
was introduced into human genetics in 1980 by Engel (1980), and in 1987 the first
case of UPD was proven by molecular methods (Créau-Goldberg et al. 1987), even
though cases of UPD had been reported earlier (for a review see Liehr 2010).
Today, more than 1,700 reports on UPD cases are available (Liehr 2011c) and what
was considered initially as something very unusual is nowadays an important
diagnostic and even prognostic factor for special syndromes (Liehr 2010) and
tumorigenesis (Tuna et al. 2009).

There are three UPD types: UPD for the entire chromosomal complement, UPD
for a complete chromosome, and segmental UPD. Additionally, two subtypes of
UPD can be recognized by molecular analysis. One is called heterodisomy (hUPD)
and is defined as inheritance of both chromosomes from one parental pair. The
other is called isodisomy (iUPD), i.e., inheritance of two copies of the same
chromosomes from one parent. hUPD and iUPD can cause disease if they affect
a gene underlying genomic imprinting (expression of a gene which depends on
parental origin). Most often the "imprinting-disorder"-related chromosomes are
tested for UPD (Online Mendelian Inheritance in Man (OMIM) 2011), i.e., paternal
UPD 6 (transient neonatal diabetes; OMIM #601410), maternal UPD 7 (Silver–Russell
syndrome, SRS; OMIM #180860), paternal UPD 11 (Beckwith–Wiedemann syndrome, BWS; OMIM #130650), maternal UPD 14 (Temple syndrome; OMIM
*605636 and #176270), paternal UPD 14 (paternal UPD 14 syndrome; OMIM
#608149), maternal UPD 15 (Prader–Willi syndrome, PWS; OMIM #176270), pater-
nal UPD 15 (Angelman syndrome, AS; OMIM #105830), maternal UPD 20 (Chudoba
et al. 1999), and paternal UPD 20 (pseudohypoparathyroidism; OMIM #103580,
#603233, and #612462). However, iUPD can further and independent of imprinting
result in functional reduction to homozygosity, and thus can cause a recessive disease
to occur in the offspring of one carrier patient. Overall, meiosis I or meiosis II errors
and/or postzygotic events contribute to UPD formation (reviewed in Liehr 2010).

sSMC and UPD are both rare events, with incidence rates of 1 in approximately
4,300 to 1 in approximately 5,000 cases. Thus, the combination of both events is
even rarer. Accordingly, only approximately 50 sSMC cases accompanied by UPD
have been reported (Liehr 2011a). Nonetheless, UPD is something cytogeneticists

have to consider as a possibility during routine sSMC diagnostics (Liehr et al. 2004; Liehr 2010). Today the following is known on UPD in sSMC cases (see Liehr 2011a, c):

- Any sSMC, irrespective of its chromosomal origin, may be principally connected with UPD, preferentially of the sSMC's sister chromosomes.
- Mixed hUPD and iUPD can be observed most often in sSMC cases, followed by complete iUPD, complete hUPD, and segmental iUPD.
- UPD of chromosomes 6, 7, 14, 15, 16, and 20 are reported most often, which may result from an ascertainment bias due to knowledge of the above-mentioned "imprinting disorders."
- Maternal UPD is approximately 9 times more frequent than paternal UPD.
- Acrocentric-derived sSMC tend to present a UPD approximately 2.5 times more often if they are mosaic with a normal cell line than corresponding nonmosaic sSMC cases.
- UPD in connection with a parentally inherited sSMC is a rare event.
- The gender of the carrier and the shape of the sSMC have no influence on UPD formation.

The question of whether UPD is a coincidence or a consequence is still a matter of discussion (Kotzot 2002; Liehr et al. 2006b).

1.4.3 Mosaicism

Somatic mosaicism is well known to be present in individuals with sSMC; about 50% of them are mosaic (Crolla 1998; Liehr et al. 2010a; Liehr 2011a).[3]

Patients with ES, CES, or i18pS tend only rarely to develop mosaicism, whereas in PKS practically every patient is mosaic at least in peripheral blood. In general, acrocentric- and non-acrocentric-derived sSMC are differently susceptible to mosaicism; non-acrocentric-derived ones are the less stable ones: 28% of acrocentric-derived sSMC and 82% of non-acrocentric-derived sSMC are mosaic. Strikingly, also in neocentric sSMC, mosaicism is more frequent in non-acrocentric-derived sSMC than in acrocentric-derived ones (58% vs. 24%). Interestingly, in TS karyotype cases with an sSMC, 76% have mosic karyotypes 46,X,+mar/45,X (Liehr et al. 2010a; Liehr 2011a).

The real grade and complexity of mosaicism seems to be even slightly higher, as recently cryptic mosaicism was repeatedly detected in sSMC cases (Liehr et al. 2006b). There were cases showing an sSMC in all metaphase spreads studied;

---

[3]Knowing that somatic mosaicism happens in approximately 50% of sSMC cases, one cannot use aCGH studies (Sect. 4.1) as a screening test to reliably detect this kind of chromosomal aberration. On the one hand, low-level mosaic cases and, on the other hand, cryptic mosaics are missed. Thus, cytogenetics is still the gold standard to detect any kind of chromosomal aberration, which then, in further steps, can be characterized by molecular (cyto-)genetic approaches.

however, interphase FISH in uncultured cells showed a mosaic situation. It is more often found that more than one variant of an sSMC is present in individual cells of a patient. At least 5% of sSMC cases have a more complex mosaicism than suggested by simple cytogenetic diagnostics. Cryptic mosaicism appears as some sSMC tend to rearrange and/or be reduced in size during karyotypic evolution. This can lead to double-ring formation or inverted duplications starting from a centric minute-shaped chromosome and in the end to formation of different variants and a highly complex mosaic, as some of the new variants can also be degraded in a subset of the cells studied (Liehr 2006, 2009b, 2011a). Interestingly, secondary rearrangements can also occur in sSMC during meiosis and hence different shapes have been reported in a mother and a daughter (Ing et al. 1987).

In the overwhelming majority of cases, somatic mosaicism has no detectable clinical effects. This seems to be due to the fact that the mosaicism rate in different human tissues is practicably not predictable and very variable (Fickelscher et al. 2007). However, there are rare cases with altered clinical outcomes due to mosaicism (Liehr 2010a). One other small group of cases occurs if sSMC appear in a structurally abnormal but balanced karyotype (McClintock mechanism; Baldwin et al. 2008; see Sect. 9.2). If in such a case mosaicism appears, i.e., loss of the sSMC, relevant genetic material is lost and this normally leads to severe clinical problems. However, if no or only very low grade mosaicism is present, the carrier of such a karyotype can be completely normal. Besides, in exceptional cases the presence of specific sSMC with known adverse prognosis was reported, but this did not lead to clinical problems, most likely due to low somatic mosaicism; examples are cases 07-W-p10/1-1, 15-O-q13/1-1, 15-O-q13/1-2, 15-O-q13/2-1, 15-O-q13/3-1, 15-O-q13.1/1-1, 22-O-q11.21/4-2, 22-O-q11.21/4-3, and 22-O-q11.21/5-1 (Liehr 2011a). Even though these are rare instances, this knowledge is extremely important for prenatal counseling.

Chapter 2
Inheritance of Small Supernumerary Marker Chromosomes

2.1 De Novo and Familial sSMC

The rate of de novo compared with familial sSMC is approximately 70% and approximately 30%, respectively (Liehr and Weise 2007). That is, more than two thirds of sSMC evolve during gametogenesis or early embryogenesis (see Chap. 3), and the remainder have been passed at least from one generation to the next. The first report of an sSMC transmitted over three generations was by Ridler et al. (1970). For de novo sSMC it is important to know that the incidence of meiotic nondisjunction increases with advanced maternal age, whereas a similar effect was not observed in males (Schinzel 2001). Thus, prenatally, sSMC are often found in pregnancies studied because of advanced maternal age (Liehr 2011a).

For familial sSMC Dalprà et al. (2005) suggested that there could be a 2:1 ratio of maternally versus paternally derived sSMC, even though this suggestion aroused controversy. However, the suggestion was confirmed (Liehr 2006), and it is now known that sSMC are predominantly inherited via the maternal line (Table 2.1). Overall, a 1.8:1 ratio is present; as detailed in Table 2.1, this quotient is more expressed in sSMC derived from nonacrocentric chromosomes. In other words, acrocentric-derived sSMC tend to be transmitted more easily from one generation to the next than nonacrocentric ones; the first can apparently survive male and female meiosis more easily, whereas nonacrocentric ones seem to face more problems there.

Familial as well as de novo sSMC can be associated with fertility problems (see Sect. 1.2.2). This observation suggests evolutionary selection against aneuploidy caused by the presence of sSMC. This might be related to the reported selection against the additional X chromosome in males with a 47,XXY karyotype (Morel et al. 2000). On the basis of the observed predominant inheritance via the maternal line (Liehr 2006), in sSMC one of the main mechanisms during spermatogenesis could be selection of gametes without an additional extra chromosome (Manvelyan et al. 2008; Oracova et al. 2009). Evidence for this taking place was provided by the fact that any kind of chromosomal aberration (including sSMC) can reduce the

Table 2.1 sSMC frequency according to chromosomal origin and parental origin (data from Liehr 2011a)

Chromosome	Cases with inherited sSMC of		Ratio
	Maternal origin	Paternal origin	
1–12; 16–20; X, Y	37	14	2.6:1
13–15; 21–22	116	71	1.6:1
Sum of all chromosomes	153	85	1.8:1

ability of correct chromosomal pairing during meiosis I, which can cause fertility problems especially in males (Shah et al. 2003). In accordance with that, oligoasthenospermia and oligozoospermia are correlated with sSMC presence (Mau-Holzmann 2005; see Sect. 1.2.2).

Besides, selection against "sperm with sSMC" could also be driven in part via fertilization success. Problems in connection with sSMC replication arising predominantly in the more rapidly progressing sperm meiosis or a "weight effect" making sperm without an sSMC more rapid than those with an sSMC, similar to the effect known from Y-chromosome-carrying versus X-chromosome-carrying sperm (Smits et al. 2005), could be envisaged as possible mechanisms involved here.

Inherited sSMC are in general harmless; however, exceptions have been reported and should be considered as a rare possibility. First, there is the problem possibly caused by sSMC formed in connection with the McClintock mechanism (Baldwin et al. 2008; see Sect. 9.2.2). Second, loss of sSMC mosaicism can negatively influence the clinical outcome of an inherited sSMC. There are cases reported in which a paternally derived, seemingly harmless sSMC caused problems, as in the patient it was present in 100% of the cells, whereas in the father it was only present in a subset of his body (Anderlid et al. 2001, case I). Also an apparently harmless sSMC present in a parent in one copy may lead to problems in the offspring when the sSMC is duplicated there (Mears et al. 1995). Finally, very rarely, secondary rearrangements have been reported in an sSMC during transmission through generations, i.e., different sSMC shapes were reported in a mother and a daughter (Ing et al. 1987).

2.2   B Chromosomes and sSMC

B chromosomes are "additional passengers found in the karyotypes of about 15% of eukaryote species. They are best understood as genome parasites exploiting the host genome because of their transmissional advantage, and are frequently not deleterious for the organism carrying them" (Camacho 2004). B chromosomes have been described for plants, fungi, insects, crustaceans, fish, amphibians, reptiles, birds, and mammals, and are present in addition to the normal chromosome content, called A chromosomes. The evolution of B chromosomes depends mainly on two factors: transmission rate (i.e., drive) and effects on fitness. For old B chromosome systems, it is plausible that they might have evolved toward neutrality (no drive or fitness effects), but it is thought unlikely that young extra chromosomes lacking drive or beneficial effects (even being neutral) might invade a population and become B chromosomes (Camacho et al. 1997).

Nonetheless, there are several similarities between sSMC and B chromosomes: both represent a heterogeneous collection of chromosomes added to the standard karyotype, both are small, both may consist of heterochromatic and/or euchromatic material (see Sect. 1.4.1), in both there is predominance of maternal transmission, and both demonstrate a tendency for mitotic instability (mosaicism; see Sect. 1.4.3). Most human sSMC seem to be evolutionary young elements, as their origin may be traced to another human chromosome through molecular analyses. Thus, according to current theories, sSMC would need drive, drift, or beneficial effects to increase in frequency in order to become B chromosomes (Liehr et al. 2008a).

Among sSMC there are at least two potential candidates that may already be or may evolve into B chromosomes: (1) sSMC stainable only by DNA derived from themselves (reviewed in Liehr et al. 2008a; see Sect. 6.25) and (2) acrocentric-derived inverted-duplication-shaped sSMC without an associated clinical phenotype (Liehr 2011a). As mentioned in Sect. 2.1, acrocentric-derived sSMC tend to be transmitted more easily through generations than nonacrocentric ones. Thus, there could possibly be a subset of familial acrocentric sSMC already behaving in a way similar to B chromosomes, and hence they could begin to spread in the population.

No definite B chromosomes have been described in humans. However, inverted duplicated derivatives of acrocentric chromosomes (especially from chromosome 15) fit the following prerequisites of B chromosome behavior: relatively high transmission rate, recurrent origin being predominantly neutral on fitness, and being on the way to a polymorphic status in the population. However, they have not yet acquired any differences in molecular nature in respect to A chromosomes. The latter is the main condition of the two cases of sSMC stainable only by DNA derived from them (see Sects. 6.25 and 7.25).

Chapter 3
Formation of Small Supernumerary Marker Chromosomes

In general, not much is known about the exact mode of sSMC formation. It is especially unclear when, why, and how during gametogenesis or embryogenesis an sSMC evolves. Nonetheless, there are models for how all kinds of sSMC shapes could be formed. These ideas are based in part on the finding that UPD and sSMC can show up together (see Sect. 1.4.2; Murmann et al. 2009) and on the observation that sSMC can evolve by incomplete trisomic rescue (Bartels et al. 2003; Stefanou and Crocker 2004). Overall, an sSMC is formed by the combination of one or more rare events happening during gametogenesis or embryogenesis.

3.1 Inverted-Duplication-Shaped sSMC

For formation of inverted-duplication-shaped sSMC, several models have been proposed (Schreck et al. 1977; Dewald 1983; Ing et al. 1987; Narahara et al. 1992). The most plausible of these is that an intrachromosomal (Murmann et al. 2008) or interchromosomal U-type exchange (Fig. 3.1) of homologous chromosomes takes place, resulting from a crossover mistake of chromatids during meiosis (Schreck et al. 1977). U-type exchange seems to be a more general mechanism of isochromosome formation, and has been found in tumor cells as well (Mukherjee et al. 1991; Wang et al. 2008).

Intrachromosomal U-type exchange was proven for neocentric inverted-duplication-shaped sSMC, where it could even be the primary mechanism of formation (Murmann et al. 2008; Sheth et al. 2009). Also, it must take place in all cases with inverted-duplication-shaped sSMC derived from the Y chromosome in TS karyotype carriers (Liehr et al. 2007b). For all other autosomal inverted-duplication-shaped sSMC no intrachromosomal U-type exchange has been yet found; it should have been expressed as iUPD. In contrast, because of different heteromorphisms present on the two centromeric and/or pericentric regions of some reported inverted-duplication-shaped sSMC, an interchromosomal origin was proven (Magenis et al. 1988; Liehr et al. 1992). Evidence was also provided that

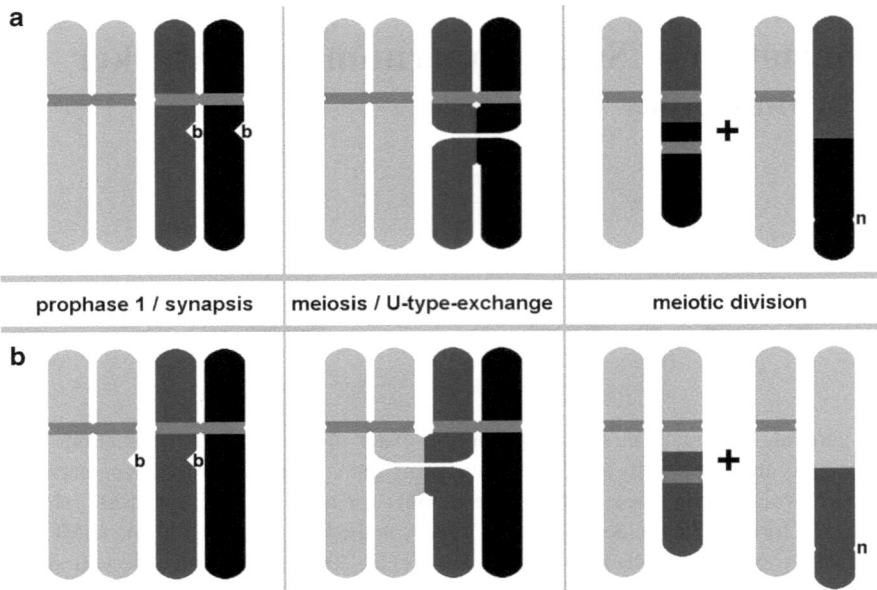

| prophase 1 / synapsis | meiosis / U-type-exchange | meiotic division |

Fig. 3.1 Inverted-duplication-shaped sSMC form by a U-type exchange mechanism. (**a**) Intrachromosomal type, (**b**) interchromosomal type. *b* break, *n* neocentromere formation can happen

this kind of U-type exchange may predominantly happen in maternal meiosis (Magenis et al. 1988; Wang et al. 2008).

N.B.: Inverted-duplication-shaped sSMC can also form after "ring opening" and inverted duplication of a centric minute, as described in Liehr (2009b) – see Sect. 3.5.

3.1.1 Centromeric Activity of Dicentric sSMC

The kinetochore is a control module located at any active centromere that both powers and regulates chromosome segregation in mitosis and meiosis. A normal human chromosome harbors one active centromere, and derivative chromosomes with two or more centromeres (dicentrics or multicentrics) are generally accepted to have only one active centromere during cell division. It is thought that several active centromeres on one piece of DNA suggest could lead to faulty alignments; two centromeres on one chromatid could orient to opposite poles, which would result in anaphase bridges and tearing of the chromosome. However, when the two centromeres are close together, there is little room for torsion between them, and stable dicentrics can be formed. Also, it has been shown that the presence of two functional kinetochores on a single chromosome does not necessarily lead to chromosome instability and loss (Sullivan and Willard 1998).

Recently, Ewers et al. (2010) studied the centromere activity in 25 dicentric sSMC. They found three different activation patterns: (1) fusion of both closely located centromeres to one active unit, (2) only one of both centromeres was active, and (3) both centromeres were active. All dicentric chromosomes studied in i18pS cases showed a functional centromeric fusion, whereas in the acrocentric-derived sSMC, in principle, all three activation patterns could be present. Centromere activity patterns of an inherited sSMC were shown to be stable over two generations.

Surprisingly, the data obtained suggest a possible influence of the sSMC carrier's gender on the implementation of the activation pattern; especially, only one active centromere was found more frequently in female carriers than in male carriers. This could still be due to an ascertainment bias and the small number of cases examined; further studies are necessary to verify that suggestion, even though the difference was statistically significant. If true, this observation might be helpful in explaining the fact that familial sSMC are transmitted predominantly via the maternal line (Liehr 2006; see Sect. 2.1).

Finally, evidence has been provided that the closer the centromeres of a dicentric are so long as they are not fused, the more likely it is that both of them will become active. In concordance with and with refinement of previous studies in dicentrics (Sullivan and Willard 1998) at a distance of 1.4 Mb up to about 13 Mb, the "two active centromere state" was favored, whereas a centromeric distance of over approximately 15 Mb led to inactivation of one centromere.

3.2 Centric Minute-Shaped sSMC

Different mechanisms of centric minute-shaped sSMC formation, including trisomic and monosomic rescue, postfertilization errors, and gamete complementation, have been proposed in the literature. Mosaicism resulting in one cell line with sSMC and one with a trisomy provided evidence for functional trisomic rescue as a real existing mechanism (Bartels et al. 2003; Stefanou and Crocker 2004). In implanted embryos the rate of trisomies was estimated to be 16% (Farfalli et al. 2007); however, no data are available on how many of them undergo trisomic rescue events. Figure 3.2 summarizes some mechanisms for how centric minute-shaped sSMC formation may occur. Some of them are correlated with iUPD and/or hUPD (Fig. 3.2a, b), and some are not (Fig. 3.2c). The models shown may account for centric (see Chaps. 5 and 6), neocentric (Liehr et al. 2007a; see Chap. 7), and complex rearranged (see Sect. 3.4 and Chap. 10) sSMC.

Even though at least some of the models depicted in Fig. 3.2 must exist, no pathways or enzymes involved in the processes of trisomic or monosomic rescue formation are known. Trisomic rescue can be the result of viable postzygotic nondisjunction or anaphase lag event occurring during early embryogenesis which can involve either trophectoderm or extraembryonic mesoderm progenitors or both of them (Kalousek et al. 1991). Los et al. (1998) added to that in 1998

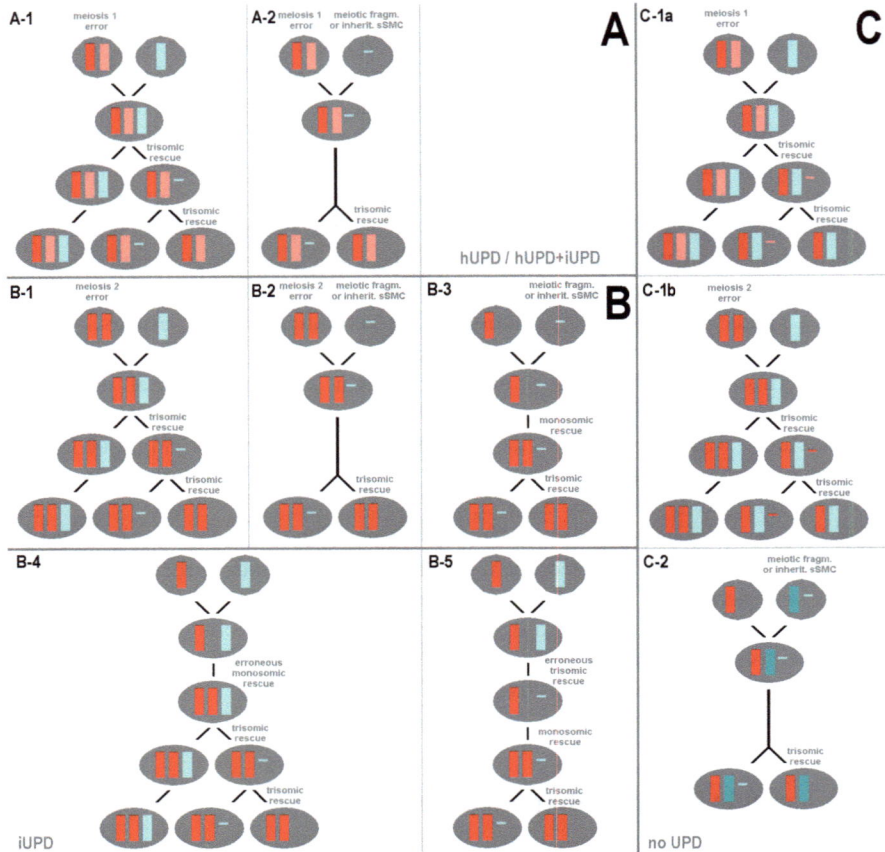

Fig. 3.2 Schematic depiction: only the sister chromosomes involved in sSMC are symbolized as *bars*. *Dark-red bars* and *light-red bars* represent the maternally derived chromosomes, *blue bars* represent the paternally derived chromosomes. Some of the possible modes of formation for minute-shaped sSMC are shown. For each example depicted, a not exclusively paternal (*blue colors*) or maternal (*red colors*) origin of sSMC is possible, but also the accordingly other variant. (**a**) sSMC in connection with heterodisomy (*hUPD*) or mixed hUPD and isodisomy (*iUPD*) can form by two mechanisms (*A-1* and *A-2*) based on meiosis I error. Variant *A-1* is that a disomic gamete forms a trisomic zygote, trisomic rescue takes place, and either a mosaic 47,XN,+A/47,XN, +mar/46,XN or a mosaic 47,XN,+mar/46,XN is formed. The second possibility is that one heterodisomic gamete meets another one, which only carries an sSMC instead of the corresponding sister chromosome. This can be either due to the presence of an inherited sSMC or due to partial chromosome fragmentation during meiosis (*A-2*). Either a nonmosaic case with karyotype 47,XN, +mar is formed, or trisomic rescue happens to the sSMC and a mosaic 47,XN,+mar/46,XN occurs. (**b**) iUPD and sSMC presence can appear because of five mechanisms (*B-1* to *B-5*). Those shown in *B-1* and *B-2* are the same as those described for hUPD before (*A-1* and *A-2*). The only difference here is that a meiosis II error led to an isodisomic zygote. Besides (*B-3*), a combination of monosomic and trisomic rescue may happen. Finally (*B-4* and *B-5*), somatic erroneous monosomic rescue followed by trisomic rescue may occur. (**c**) sSMC without uniparental disomy (*UPD*) may arise after meiosis I or meiosis II errors similar to *A-1/B-1* and *A-2/B-2* (*C-1a* and *C-1b*). Model *C-2* shows how inherited sSMC can be passed through generations without UPD

the theory of chromosome demolition as an alternative correction mode to trisomic rescue. In sSMC formation, chromosome demolition would be a process of deliberate fragmentation and/or removal of one of the sets of three chromosomes during anaphase or metaphase. Such chromosome fragmentation is seen in Howell–Jolly bodies (Felka et al. 2007) and a case with a del(5)(q31) was also interpreted as incomplete chromosome fragmentation (Vialard et al. 2009), too. Also, the recently recognized phenomenon that "developmental chromosome instability" is significantly increased during the embryonic stage and affects different tissues should be mentioned in this context (Iourov et al. 2009). Los et al. (1998) "consider trisomic rescue to consist of one correction event in the first to fourth postzygotic cell division with a subsequent unknown distribution of trisomic and disomic cells among the progenitor cells of the inner cell mass and trophoblast compartment until 16-cell stage." Cellular selection during the following formation of placenta and early embryogenesis would help, as a result, to ensure the presence of a numerically balanced chromosome complement in the developing fetus.

Together with the findings that there are inherent epigenetic differences between the paternal and maternal pronuclei in early cleavage stage embryos (Wu and Chu 2008) this led to the following idea, in part based on the 1:9 rate of paternal UPD to maternal UPD in sSMC mentioned in Sect. 1.4.2. As well as the fact that aneusomies are more likely to be contributed from the female side (Bán et al. 2003), another kind of enzymatic content in male- and female-derived pronuclear compartments could also be important. The oocyte has a less active machinery to eliminate chromosomal mistakes than the spermatocyte. Thus, at the pronuclei stage, an elimination of a paternally derived additional chromosome could be more likely than that of a maternally derived one (Liehr 2010). In concordance with this, evidence for the existence of a chromosome counting mechanism in the zygote and early embryogenesis has been provided (Migeon et al. 1996). Also, the recently discovered "chromosome kissing" could be involved here (Augui et al. 2007).

N.B.: Centric minute-shaped sSMC can also form by "ring opening" as described in Liehr (2009a) – see Sect. 3.5.

3.3   Ring-Shaped sSMC

Several possibilities of how ring-shaped sSMC may evolve have been proposed. First, such a kind of sSMC can be formed in association with a deletion of part of the chromosome. This leads to a balanced situation in the carrier and is known as the McClintock mechanism (Baldwin et al. 2008; Fig. 3.3, panels A-1, A2). Clinical problems only arise if exclusively the sSMC or the derivative chromsome from which the sSMC was derived are passed to a carrier's child, as then a chromosomal imbalance – either partial trisomy or partial monosomy – is present. In such ring-shaped sSMC, parts of the centromere can be included, leaving two centric chromosome fragments, one of which forms a small ring (Fig. 3.3, panel A-1), or a neocentromere is formed within the ring-shaped sSMC (Fig. 3.3, panel A-2; Liehr et al. 2007a) or the derivative.

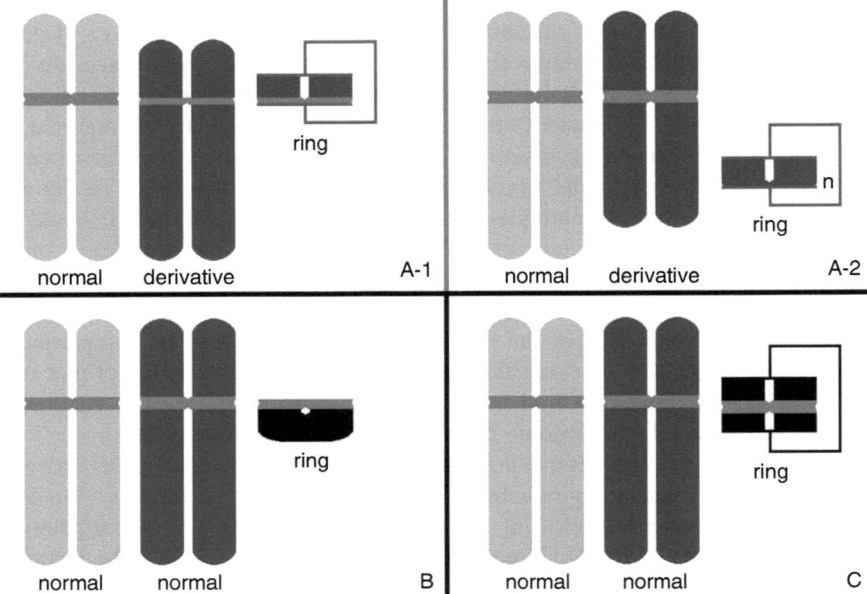

Fig. 3.3 A ring-shaped sSMC can form as follows. In a balanced karyotype parts of the sSMC's sister chromosome are excised and stabilized by ring formation (McClintock mechanism). Either the ring-shaped sSMC and the derivative chromosome share the centromeric region (*A-1*) or a neocentromere (*n*) is formed on the sSMC (*A-2*) or the derivative (not depicted). Ring formation can be due to an intrachromosomal U-type exchange (*B*, Fig. 3.1). A ring-shaped sSMC can evolve from a centric minute-shaped sSMC (*C*)

Second, ring formation has been proposed in connection with an inverted duplication as due to a U-type reunion between broken sister chromatids (Michalski et al. 1993). This kind of ring has only rarely been reported for sSMC, and even then was observed in "larger" sSMC or SMC (Starke et al. 2003). This might be connected with steric problems this mechanism may face in sSMC (Fig. 3.3B).

Third, for the overwhelming majority of ring-shaped sSMC a ring formation starting from a centric minute is suggested, which during karyotypic evolution acquires the ring shape, maybe to become more stable (Liehr et al. 2004; Fig. 3.3C).

N.B.: The formation of double rings is well known and frequently observed. It is thought to be due to a sister chromatid exchange with a normal centromere division (Ramirez-Duenas and Gonzalez 1992) – see Sect. 3.5.

3.4 Complex sSMC

Complex rearranged sSMC (Trifonov et al. 2008) are only identifiable as such after molecular (cytogenetic) analysis. In cytogenetic analysis they look like centric minute-shaped, ring-shaped, or inverted-duplication-shaped sSMC. Most complex rearranged sSMC are represented by the cases with ES (Carter et al. 2009; see Sect. 5.1). The carriers of this special derivative chromosome 22 (der(22)t(11;22) (q23;q11)) normally inherit it from a parent who has a balanced translocation t(11;22)(q23;q11). An embryo with the karyotype 46,XN,der(22)t(11;22)(q23; q11) is not viable. Thus, patients with ES have had to double their only chromosome 22 during early embryogenesis (see Fig. 3.2, panel B-3), or gamete complementation must have taken place (see Figs. 3.2, panels A-2, B-2, and C-2).

Complex sSMC besides those in ES (for a review see Trifonov et al. 2008) derive either from a single chromosome (Fang et al. 1995; Stavropoulou et al. 1998; Stankiewicz et al. 2001), from two chromosomes, or even from three different chromosomes (for a review see Liehr et al. 2004). Models for how they form are not available yet, even though copy number variant regions are thought to be causative for rearrangements (Zhang et al. 2009; Mefford and Eichler 2009; see Sect. 3.7).

3.5 Mixtures of Different Shapes

It was recently observed that one, two, or all three sSMC shapes (minute, ring, inverted duplication) can be present in a single patient with karyotype 47,XN,+mar (Liehr et al. 2006b). As in such patients initially only one sSMC type was present in the early embryo or zygote, karyotypic evolution must have taken place subsequently. If different sSMC shapes are present, which often can only be characterized in detail by molecular cytogenetics, this condition is called cryptic mosaicism. Several patients with extremely active karyotypic evolution have been reported with up to ten different sSMC variants of the same derivative chromosome in their peripheral blood cells (Liehr et al. 2006b; Liehr 2009b). Examples of how different shapes of sSMC can change to other ones are given in Fig. 3.4. Presently, it can just be stated that this flexibility in sSMC shape exists; there are as yet no ideas on the mechanism of ring formation from a minute-shaped sSMC, for ring doubling, ring opening, and formation of inverted-duplication-shaped sSMC from centric minute-shaped sSMC, or for reduction of sSMC size and subsequent stabilizing of the sSMC again.

Note that in TS karyotype cases sSMC of all three shapes also appear. However, their formation has not been studied yet nor is it understood. Also, there are TS cases with larger, differently shaped X-derived derivative chromosomes (Liehr et al. 2007b).

3.6 Multiple sSMC

As outlined before, multiple sSMC derive from a different chromosomal subset from single sSMC (see Sect. 1.3.3; Fig. 1.3). Besides, they have a different distribution of shapes from centric sSMC in general. Whereas in single sSMC there is a difference in the shape distribution distinguishing acrocentric-derived sSMC from

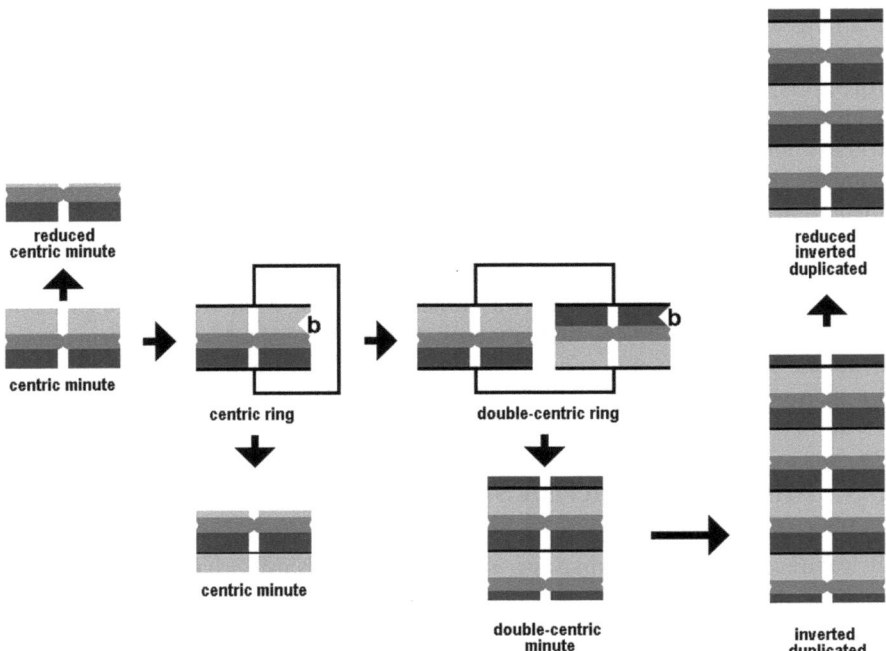

Fig. 3.4 Multiple shapes of sSMC can evolve during the lifetime of an sSMC carrier. In the schematically given example according to the case reported in Liehr (2009a), it is postulated that the starting point is a centric minute-shaped sSMC. This can undergo ring formation (*short horizontal arrows*), reduction in size (*vertical arrows to top*), ring opening (*vertical arrows to bottom*), and inverted duplication (*long horizontal arrow*)

non-acrocentric-derived ones, this is not the case in multiple sSMC (Fig. 3.5). Also, centric minute-shaped ones are most frequent in multiple sSMC followed by ring-shaped ones and inverted-duplication-shaped ones. Beverstock et al. (2003) suggested correctly that the formation of multiple sSMC of different chromosomal origin is based on a mechanism different from the mechanisms discussed above for single sSMC. Daniel and Malafiej (2003) proposed multiple sSMC may originate from transfection of chromosomes into the zygote derived from one or more superfluous haploid pronuclei that would normally be degraded. Also, rescue of a triploid zygote could be the reason for multiple sSMC (Liehr et al. 2011). However, no studies are available supporting any of these ideas.

3.7 Breakpoint Characteristics of sSMC

Copy number variant regions are thought to be causative at least for a certain number of chromosomal rearrangements (Zhang et al. 2009; Mefford and Eichler 2009). Accordingly, 124 sSMC breakpoints located in euchromatic chromosomal

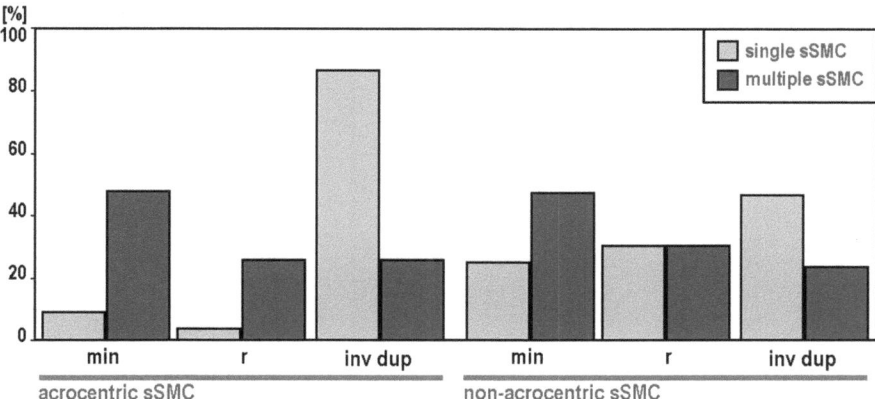

Fig. 3.5 Frequency of shapes of sSMC in single sSMC compared with multiple sSMC

regions were characterized on a molecular level and analyzed for common features using the published human genome sequence as a reference. It was confirmed that more than 99% of the sSMC breakpoints are located within copy number variant regions and/or segmental duplications. Moreover, approximately 75% of the breakpoints were concordant with so-called fragile sites of the human genome (for a review see Mrasek et al. 2010). Still, there was an approximately 10% overlap of the observed breakpoints and interspersed telomeric sequences (unpublished data). Thus, sSMC do not break and/or stabilize at certain end points by chance; instead they break and/or stabilize within somehow predestined regions. This finding should facilitate the characterization of clinically relevant critical regions in sSMC (see Chap. 6).

Chapter 4
Small Supernumerary Marker Chromosomes in Genetic Diagnostics and Counseling

Generally, sSMC are detected in four groups of patients: (1) prenatally studied ones (with and without sonographic abnormalities), (2) postnatally examined adults with fertility problems, (3) children and adults with unexplained mental retardation, developmental delay, and/or dysmorphism, and (4) patients in whom sSMC can be an incidental finding when cytogenetic analysis is performed for other reasons (see Sect. 1.2). Because of their relative rarity, and because it is impossible to do a directed search for a person with sSMC, the detection of such an additional marker is always surprising and unexpected for the cytogenetic laboratory performing the analysis.

4.1 sSMC Diagnostics

4.1.1 First Steps: Banding Cytogenetics

Classically, the presence of an sSMC in a patient's peripheral blood, amniocytic, or chorion-derived cells is substantiated by chromosome banding analysis (Seabright 1971; see Chap. 1). In Fig. 4.1, typical examples of sSMC are shown; the shape and size can vary in the three basic shapes.

Nowadays, some laboratories are starting to skip the chromosome analysis and perform DNA-chip-based methods such as aCGH first, or even as the only test (Tabor and Cho 2007). Thus, sometimes the first hint on an sSMC can come from aCGH (Ahn et al. 2010). Because this kind of approach only detects a chromosomal imbalance but gives no information of whether it is due to an sSMC, a duplication, an insertion, or an unbalanced translocation, a karyotype/chromosome analysis has to be done additionally. It is important to do this because each of the chromosomal rearrangements causing an imbalance mentioned have other impacts on heritability, and thus on genetic counseling. Furthermore, cytogenetic and/or molecular cytogenetic studies should be done; as in some cases because of mosaicism a partial tetrasomy might be interpreted as partial

T. Liehr, *Small Supernumerary Marker Chromosomes (sSMC)*,
DOI 10.1007/978-3-642-20766-2_4, © Springer-Verlag Berlin Heidelberg 2012

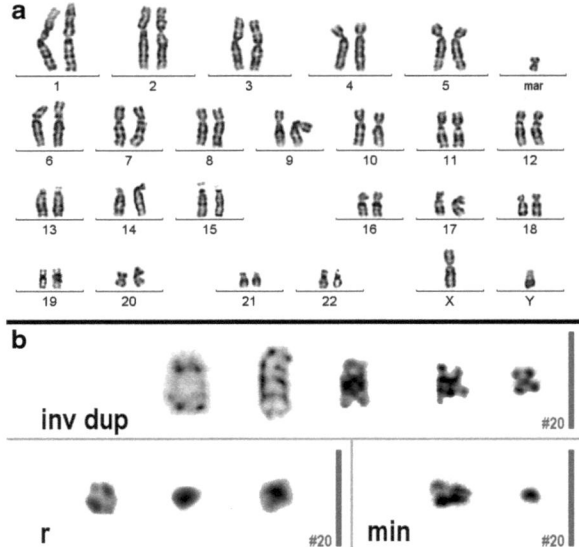

Fig. 4.1 (**a**) Karyogram of a typical sSMC found in prenatal diagnostics; karyotype 47,XY,+mar. Later the carrier turned out to suffer from isochromosome 18p syndrome. (**b**) sSMC as detected in karyotyping after G bands by trypsin using Giemsa (GTG) banding. Five examples are shown for inverted-duplication-shaped sSMC (*inv dup*), three for ring chromosome type (*r*), and two for centric minute-shaped ones (*min*). As a reference the size of chromosome 20 of the corresponding metaphase spreads is given as a *gray bar labeled with #20*

trisomy (Liehr 2011a, case 18-Wi-156), or because of DNA homologies of the short arms of the acrocentric chromosomes and the pericentric region of chromosome 9 (Starke et al. 2002) a wrong chromosomal origin may be suggested (Liehr 2011a; 15-O-q11.2/2-1).

When an sSMC is detected by banding cytogenetics, its approximate mosaic state can be determined at the same time. However, to define the real rate of mosaicism in a certain tissue, uncultured tissue samples have to be used. This is necessary, as in lymphocyte culture cells with and without sSMC may proliferate differently from those in the human body (Fickelscher et al. 2007).

Furthermore, two important studies can be performed by a cytogenetic laboratory to solve some of the problems correlated with an sSMC detection (Table 4.1):

1. By application of silver staining [nucleolar organizing region (NOR) staining; Bloom and Goodpasture 1975] and C-banding [C bands by barium oxide using Giemsa stain (CBG) staining; Pardue and Gall 1970], the size of the heterochromatic proportion of the sSMC can be approximately determined approximately. Also, acrocentric-derived and non-acrocentric-derived sSMC can be distinguished to a good extent.
2. A chromosome analysis of the sSMC carrier's ancestors can solve the question of the parental origin of the sSMC detected.

Table 4.1 Seven questions to be answered when an sSMC is detected and to what extent they can be answered by applying the approaches listed

Questions	Answer by				
	GTG	NOR/CBG	FISH	aCGH	test UPD
Q1. Is there an sSMC?	+	+	+	(+)	(+)
Q2. Is the sSMC de novo or parental?	+	+	+	+	−
Q3. Is the sSMC present in a mosaic?	+	+	+	(+)	−
Q4. Where does the sSMC originate from?	−	(+)	+	(+)	(+)
Q5. Is there euchromatin on the sSMC?	−	(+)	+	+	(+)
Q6. Is there cryptic mosaicism?	−	(+)	+	−	−
Q7. Is there a UPD in connection with the sSMC?	−	(+)	(+)	(+)	+

GTG G bands by trypsin using Giemsa stain, *NOR* nucleolar organizing region staining, *CBG* C bands by barium oxide using Giemsa stain, *FISH* fluorescence in situ hybridization, *aCGH* array-based comparative genomic hybridization, *UPD* uniparental disomy test, + question can be answered by this approach, − question cannot be answered by this approach, (+) in some instances this question might be answered (in parts) by this approach

In Table 4.1 the possibilities of the different techniques of characterizing an sSMC discussed in this section are summarized. Questions Q1–Q3 (Table 4.1) can be answered by GTG banding, NOR staining, and CBG staining. In some cases, the latter two approaches may provide some hints as to how questions Q4–Q7 should be answered.

4.1.2 Molecular Cytogenetics

If cytogenetic analysis has detected the presence of an sSMC, its further molecular cytogenetic characterization is driven by three main questions (Q4–Q6 in Table 4.1).

In specialized laboratories a comprehensive marker chromosome characterization can be done by various multicolor FISH approaches (Liehr 2009a, 2011b) which, in the end, answer these three questions, sometimes even the question related to UPD presence (Q7; Weise et al. 2008b). The particular techniques used may be microdissection and reverse FISH (Fig. 4.2a; Starke et al. 2001), multicolor FISH applying all 24 human whole chromosome painting probes (Fig. 4.2b; i.e., multiplex FISH, Speicher et al. 1996; spectral karyotyping, Schröck et al. 1996), (sub)centromere-specific multicolor FISH (Fig. 4.3a, b; Nietzel et al. 2001; Trifonov et al. 2003; Liehr et al. 2006a), FISH-based chromosome banding such as multicolor banding (Fig. 4.3c; Weise et al. 2008a), and even aCGH (Fig. 4.2a; Backx et al. 2007; Pietrzak et al. 2007; Tsuchiya et al. 2008).

Overall, applying FISH approaches, one can characterize an sSMC almost comprehensively (Table 4.1). Only for the detection and characterization of a UPD is microsatellite analysis or a methylation test still a favorable system (see Sect. 4.1.3). aCGH is the most recent molecular cytogenetic way to characterize

Fig. 4.2 (a) Metaphase spread with an sSMC derived from chromosome 2 as characterized by reverse fluorescence in situ hybridization (FISH) using a microdissection-derived sSMC-specific DNA (*midi-probe*) as a probe. Specific red signals are obtained on the sSMC (*mar*) and both chromosomes 2 (*#2*). In the *lower-right corner* the array-based comparative genomic hybridization result is shown proving partial trisomy 2q (*arrowhead*) induced by the sSMC. (**b**) A neocentric sSMC derived from chromosome 13 (*mar*) as characterized by multiplex FISH

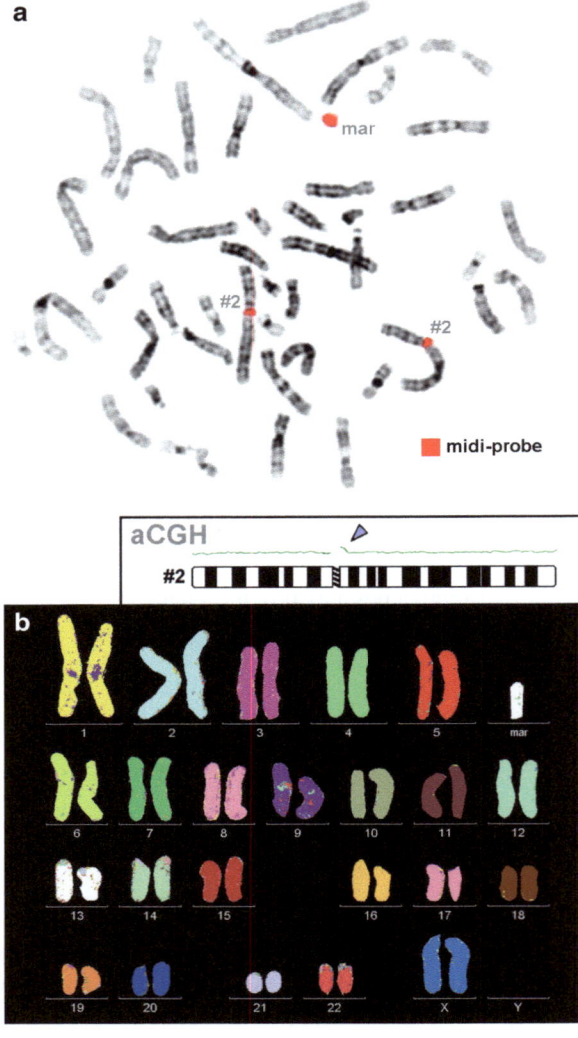

chromosomal aberrations. However, as shown in Table 4.1, aCGH alone is not able to answer the question regarding cryptic mosaicism (Q6), may have problems finding an sSMC at all, especially if it is present in a mosaic state (Q1, Q3), may have problems characterizing even the sSMC's chromosomal origin if it is exclusively heterochromatic (Q4), and can only answer the question regarding UPD in the sSMC's sister chromosomes (Q7) if a special kind of aCGH is used, called single nucleotide polymorphism aCGH (Zhou et al. 2005). If an sSMC is of parental origin, question Q2 can be studied by aCGH; however, it is quite expensive to answer this simple question in this way. aCGH has real

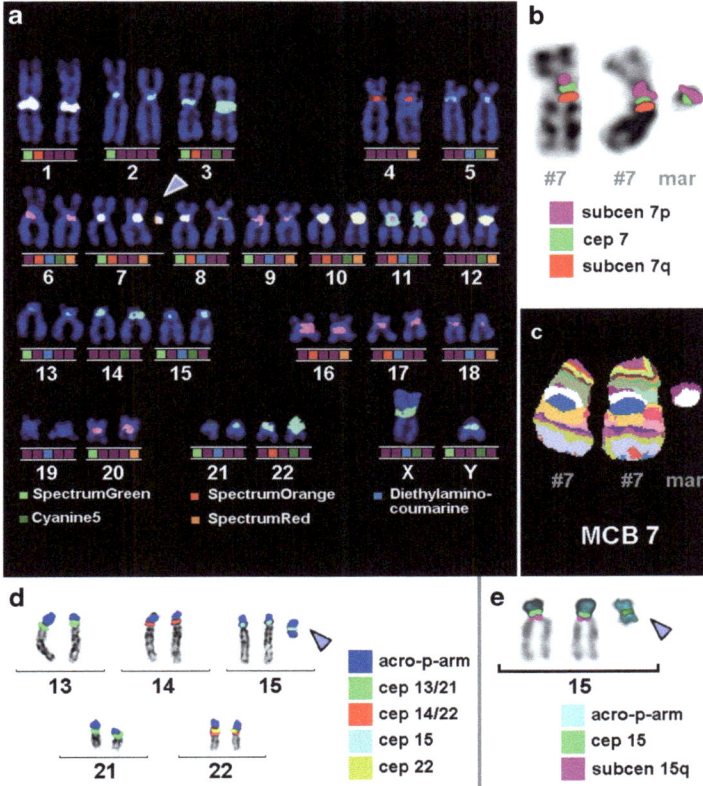

Fig. 4.3 (**a**) Karyogram after centromere-specific multicolor FISH (cenM-FISH) identifying an sSMC as derived from chromosome 7 (*arrowhead*). (**b**) By subcentromere-specific multicolor FISH (subcenM-FISH) for chromosome 7 the presence of euchromatic material on the sSMC from Fig. 4.3a could be proven. This was exclusively derived from the short arm. (**c**) Multicolor banding confirmed the results obtained by cenM-FISH and subcenM-FISH. (**d**) Partial karyogram exclusively showing the acrocentric chromosomes of another sSMC case. Here the derivative was derived from chromosome 15 and showed an inverted duplicated shape by a probe set called acrocentric-chromosomes-directed centromere-specific multicolor FISH. (**e**) subcenM-FISH showed in the sSMC depicted before in Fig. 4.3d the absence of euchromatic material. Thus, the sSMC consists exclusively of heterochromatin. *acro-p-arm* probe specific for the short arms of all acrocentric chromosomes, *cep* centromeric probe, *subcen* centromere-near probe, *mar* sSMC

advantages in sSMC diagnostics and research owing to its ability to characterize the exact size of the induced imbalance in a single experiment (Q5, Fig. 4.2a; see Sect. 3.7).

Luckily, sSMC cases can also be solved using simpler FISH tests than those mentioned above (Figs. 4.2 and 4.3). A flowchart for how sSMC can be studied if routine FISH approaches are at hand is given in Fig. 4.4.

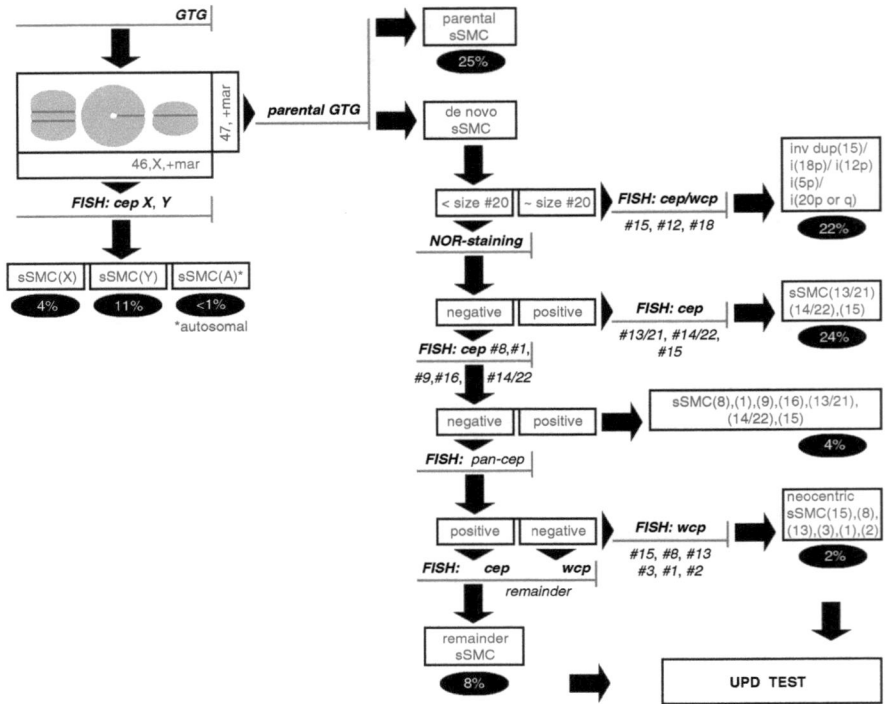

Fig. 4.4 Flowchart showing how to characterize a single sSMC. The methods and probes applied are written in *black and italics*. The results obtained are in *gray in black squares*. The approximate percentage of sSMC expected to be characterized after each step is indicated in *white letters in black ovals*. sSMC present in Turner syndrome karyotypes and in karyotypes of 47,XN,+mar are included in this scheme. Thus, the percentage given for parentally derived markers, normally referring only to the latter group, is indicated here lower than in the text. *cep* centromeric probe, *pan-cep* pan-centromeric probe, *wcp* whole chromosome painting probe, *NOR* nucleolar organizing region

4.1.3 Molecular Genetics

As already mentioned, after identification of the origin of the sSMC, its normal sister chromosomes should be tested for their parental origin to exclude a possible UPD (see Sect. 1.4.2). UPD can be studied by molecular genetic approaches, such as microsatellite analysis (von Eggeling et al. 2002; for the method see Weber 1990) or methylation-specific polymerase chain reaction (PCR) (Nietzel et al. 2003; for the method see Fraga and Esteller 2002). Recently, parental origin determination FISH was reported as a molecular cytogenetic alternative to determine UPD at a single-cell level (Weise et al. 2008b).

Questions Q1, Q4, and Q5 in Table 4.1 might also be partly answered by microsatellite analysis to check for UPD of the sSMC's sister chromosomes. If the sSMC is euchromatic and an informative microsatellite marker is in the trisomic region, even the parental origin of the sSMC itself might be determined by that test.

As well as a UPD test, tests for chromosomal imbalances can be done in some instances by molecular genetic approaches such as multiplex ligation-dependent probe amplification assay. This is a PCR-based test system which can quickly and at a relatively low cost detect the typical gain of copy number in, e.g., ES (Medne et al. 2010).

4.1.4 How To Characterize an sSMC

A scheme for how sSMC could be best characterized in a routine (molecular) cytogenetic laboratory was recently described (Liehr et al. 2009) and is given here in a slightly modified form (Fig. 4.4).

Step 1. If karyotype 45,X/46,X,+mar correlated principally to TS is found in GTG banding, FISH with a centromeric probe specific for X and Y chromosomes has to be done. In more than 99.5% of these cases, the sSMC will be characterized. If an sSMC derived from the X chromosome was identified, the commercially available XIST probe, specific for the X chromosome inactivation region, may be helpful for prediction of the clinical outcome. If the sSMC is derived from the Y chromosome the risk of gonadoblastoma has to be considered (see Sect. 5.5). In the case of karyotype (mos) 47,XX,+mar or 47,XY,+mar, go to step 2.

Step 2. Perform GTG-banding analysis of parental peripheral blood to define if the sSMC is de novo. In prenatal cases with limited time, the characterization of the marker and the parental cytogenetic analysis should be done in parallel. If the sSMC is inherited, go to step 3; if it is de novo or if its parental origin cannot be determined, go to step 4.

Step 3. For inherited sSMC the identification of the origin of the sSMC may be replaced by genetic counseling and by closely monitoring the pregnancy by high-resolution ultrasonography. However, in exceptional cases, an inherited sSMC can be connected with clinical abnormalities (Baldwin et al., 2008). If the origin of the sSMC needs to be clarified nonetheless, which is always to be preferred, go to step 4 by analyzing the corresponding parental blood sample.

Step 4. If the sSMC has almost the size of chromosome 20 of the same metaphase spread, the presence of a large inverted duplication chromosome 15 (inv dup(15)), an isochromosome 18 (i(18p)), or an isochromosome 12 (i(12p)) should be excluded first by applying the corresponding centromeric and/or whole chromosome painting probes. When the sSMC's origin has been clarified, go to step 10. If the sSMC's origin has not been determined, chromosomes 13, 9, 5, and 20 should be tested; if they also test negative, go to step 5.

Step 5. If a clear positive NOR silver staining is present on the sSMC, the latter's origin can be determined by hybridizing commercially available centromeric probes for all acrocentric chromosomes, i.e., chromosomes 13/21, 14/22, and 15. If the sSMC's origin has been determined, go to step 10; otherwise go to step 6.

Step 6. Use commercially available centromeric probes, testing sequentially for chromosomes 8, 1, 9, and 16, according to Fig. 1.2 the most frequent non-acrocentric-derived sSMC. Then, if the sSMC has tested NOR-negative, test for chromosomes 14/22, 15, and 13/21, as there are cases reported with sSMC derived from acrocentric chromosomes, but without NOR. If the sSMC's origin has been clarified, go to step 10; if it has not been clarified, go to step 7.

Step 7. To determine if the case studied is one of the rare cases with a neocentric sSMC, a commercially available pan-centromeric probe should be applied. This test has to be performed, as neocentric sSMC nearly always have a clinically adverse prognosis (Liehr et al. 2007a). In approximately 3% of cases no alpha-satellite DNA is present on the sSMC. If the sSMC has alpha-satellite DNA, go to step 8; if the sSMC has no alpha-satellite DNA, go to step 9.

Step 8. The sSMC with alpha-satellite DNA can derive from 11 different human chromosomes. If there is enough material to continue the analysis, proceed in the following sequence (applying centromeric probes if nothing else is mentioned): chromosome 19 (whole chromosome painting probe), chromosome 2, chromosome 3, X chromosome, chromosome 17, chromosome 7, chromosome 4, chromosome 6, chromosome 11, chromosome 10, and Y chromosome. Then go to step 10.

Step 9. To characterize the origin of a neocentric sSMC one has to perform FISH applying whole chromosome painting probes in the following sequence: chromosomes 15, 8, 13, 3, 1 and 12. In approximately 75% of cases the neocentric sSMC's origin will clarified by now; if it has not been clarified, go to step 10.

Step 10. In 5–10% of de novo sSMC cases, a UPD of the sSMC's sister chromosome is described (Liehr 2010); thus, a UPD test should be considered.

Using this scheme, one can characterize up to 92% of sSMC at least for chromosomal origin. Further characterization can be done in specialized laboratories, such as that of the author.

4.2 Personal Experiences of Patients Receiving the Diagnosis sSMC Prenatally

In the Preface a moving story provided by an intentionally anonymous family from somewhere in Europe was reported. A pregnancy was terminated because of the presence of an sSMC in the fetus which later turned out to be harmless and a familial variant. For this book, for the first time families facing the diagnosis of sSMC were kindly asked to tell their stories. These reports are given throughout the book at appropriate positions. Further examples of personal experiences of families living with children with an sSMC leading to clinical problems can be found at http://www.rarechromo.org.

4.2.1 Personal Experience 1

Listed in Liehr (2011a) *as 15-O-q11.2approximately 12/1-2; reported by the mother of the now 6-year-old girl.*

Six years ago, we already had a 4-year-old daughter and longed to have a second child. Logic, that we were very happy when we learned of our second pregnancy. I was already 39 years old, so I did a chorion biopsy [chorionic villus sampling (CVS)] in early pregnancy, as I had also done 4 years before in my first pregnancy. Thus, I knew the result should be there after few days. When I had no information after 1 week I became troubled and called the human genetic center. I was told the MD would call me back and I became more anxious. The phone call came really fast and I was told carefully that an sSMC had been found. At that moment I had the feeling to fall into a deep hole. That what was carefully, kindly and comfortingly said to me by the genetic counselor had no chance to reach me. After the call I sat bewildered in my living room and could only cry – a chromosome disorder, twig light zone big, further blood tests necessary from father and mother,. . . . I had thousands of questions in my mind. For me as a brain-driven person the CVS has only been one step for a final confirmation that everything was o.k. with the developing child! But now nothing was o.k. anymore.

The next day I had to go on a business trip abroad and could not deliver my blood for the parental test. These days were awful; in the evening I checked the internet for information. One thing was clear to me: sSMC are not completely unusual, and even though 70% of persons with an sSMC are clinically normal there are these 30% which have problems. And these problems are in a twig light zone, as there are still many sSMC which are not clearly studied and without published reference cases.

In the meantime my husband and I had given our blood for parental test and I had arrived at a point where I hoped everything would be turning out all right. However, we were also thinking about a bad outcome. A positive surprise was that in such a situation one can find support by others, by people who went through similar things (see Appendix) *or who are specialists in that field. I would have been opting for the child irrespective of blood test result, as all sonographic results were normal. Certainly, I do not believe that one can order healthy children from God or from somewhere. There are no guarantees.*

The results from the genetic laboratories said that I am also carrier of the same sSMC which was also found in my developing child (47,XX,+inv dup(15) (q11.2approximately 12)). As I am clinically normal it could be expected with high probability that our child would also be healthy born. For completeness I want to add that also a UPD 15 was excluded.

In 2006, our daughter was born by caesarian section and we were overjoyed when we had our little girl in our arms. She was and is happy, full of energy and has her parents and her big sister completely under control.

We learned from this pregnancy that there is no guarantee for anything in life and that the twig light zone in prenatal diagnostics still may be large. Also, we became more attentive and sensible to many things. Maybe it sounds pathetic, however, we got a little more humility and do not take many things for certain any more.

4.2.2   Personal Experience 2

Listed in Liehr (2011a) as 22-Wces-5-117; reported by the father of the now 2-year-old girl, Czech Republic.

My wife and myself live in the Czech Republic. In summer 2008 *we married and shortly after honey moon my wife got pregnant. We were very happy about that news as this meant for us a step into our future as a family, and as it was our first child, which we wished to have a lot. All went well, without any complications, the way most women wished to have a pregnancy – no attacks of sickness or any other problems.*

According to diverse controls at the gynecologist my wife and child developed exactly as expected according to the percentiles. Thus, we had no reason to be worried.

In the 16th week of gestation, the AFP [alpha fetoprotein] test detected a severely enhanced risk for Down syndrome of 1/100. Thus, the gynecologist assigned us for chromosome analysis in a specialized genetic laboratory in Plzen (about 2 h to drive by car from us). So in the same week amniocentesis was performed. And we also took the chance for a 3D sonography. We obtained the first photos and the proof to expect a little girl. Still the pregnancy developed normal and for us everything was OK. The result of amniocentesis was expected for week 20; for week 21 a more detailed sonography was advised.

In week 20 of gestation (5 days prior to Christmas eve) the result of amniocentesis was presented to my wife as: karyotype 47,XX,+mar[29]. Due to that my wife called me and told me there was something wrong. She could not give me any details of the MD as she was too agitated. Thus, I called him and he told me that there was no reason to be worried, but the baby had one chromosome more than normal. Due to Christmas we were asked to come to Plzen in the 21st week of gestation (2 days prior to New Year's Eve) to have some more explanations in a personal consultation.

In the 21st week of gestation we were informed that the baby had an additional part of a chromosome, but no Down syndrome, however, the origin of the chromosome was unknown and could not be determined. As the counselor told us the risk for clinical problems due to the marker chromosome was approximately 10% we decided to continue the pregnancy. To further minimize the risk we were asked to let study our own blood cytogenetically. If one of us had had the same marker, the risk of the baby would have been near zero, as both of us are completely healthy.

Along the entire consultation I had the feeling that something was wrong, which was not told by the MD to us, he did not dare to tell us. For further requests we obtained his private phone, as New Year was coming. Afterwards we went to Plzen for checking the baby's heart by ultrasound; diagnosis: everything OK.

In the 23rd week of gestation we got the result of cytogenetics for my wife and me; none of us had the additional chromosome. While talking to the MD I had the impression that he would have liked to tell us we should terminate the pregnancy, however, he could not do for ethical reasons.

Not knowing what to do, we started to do some own research in the internet to be able to better understand the situation we were in now. After only 30 min of search in Google we found reports on sSMC in the Institute of Human Genetics Jena. Especially the research of Dr. Liehr gave us new hope. There were methods to find out where the sSMC comes from, what it consists of and what symptoms might be correlated with.

Now we started a race against the time. My wife was already in week 23 of her gestation. Next day I contacted Dr. Liehr by telephone, telling him our story. Quickly we had the idea to send the material by express mail from the laboratory in Plzen to Jena to obtain at least the maximum of information. The laboratory in Plzen was extremely co-operative. However, in the end it turned out to be the best if I personally picked up the probes in Plzen and brought them to Jena. The same day my wife and I delivered the material to Jena and returned home after having an informative talk with Dr. Liehr on our problem. Already the next day (a Friday) we had the information that the sSMC was derived from chromosome #22. Due to the weekend, the more detailed analysis of the sSMC had to wait until Monday. With this result began the second part of our research in the internet, to answer the question which symptoms can be correlated with an sSMC derived from chromosome #22. The more we read the clearer became to us that Down syndrome is not the worse one can have to face. More and more we expected that our child would have the cat eye syndrome, which apart from physical also mental and motor defects could be expected, which are not to be detected or excluded by ultrasound.

Together with my wife I defined the clear rule for a decision for an induced termination. Even though this was not easy, we tried to do it based on scientific rather than on emotional grounds. Here the research results in sSMC nowadays *available were extremely helpful. The whole weekend we cried a lot and isolated ourselves. We were in a situation "we would not wish our biggest enemy" as one says in German. The emotions went up and down, according to the article we read in internet.*

In parallel, we were in contact with the MD in Plzen, as my wife was in week 24 of gestation and we had to prepare everything for the worst case. Already on Friday we had reserved a place in Plzen hospital. On Tuesday at around 12 a.m. the result from Jena came: 47,XX,+mar[100%] de novo .ish inv dup(22)(q11.21), i.e., the developing child would have a cat eye syndrome, as apprehended somehow by us.

Thus, for both of us it was clear that the unborn child would never see the light of the world. The worst night of our life was in front of us, especially for my wife, who already felt the baby's movements in her body. We felt like carrying our child to the executioner, but we were aware of the fact that in our today's world life was already difficult for healthy children, not to talk about diseased ones.

Next day (Wednesday) we went to Plzen in the genetic laboratory for a final screening. In ultrasound everything was OK, weight and size were as expected, but the mental and motor development of the little baby could not be visualized. The remark on the part of the MD "I would have made the same decision" was only a little solace. After termination, which was not easy, and lasted 2 days, an autopsy was performed. The result was a flat profile in the face of the baby and otherwise normal physical parameters.

Overall, the enervating race against time, the late abortion, the extreme emotional and psychological pressure, the many questions ("why"), the difficulties to find a decision would not have been necessary if the maximal possible information had been available to us already in an earlier phase of the pregnancy. We were privileged as we had access to internet and the possibility to communicate in German and English. Not all parents have that possibility and their decisions are based on partial information. Here we want to thank especially Dr. Liehr and his team for the possibility to find the right decision. Also, we thank all parents who posted their reports in the internet and the MDs and coworkers in the genetic laboratory in Plzen and Plzen hospital. I thank my wife for her mental and psychic strength in a situation which was harder to carry for her than for me.

There is a little consolation we have in this sad story. For the next child there is no risk for repetition.

Short follow-up report: *In summer 2010 our first healthy daughter was born and we are very happy and proud of her.*

4.2.3 Personal Experience 3

Listed in Liehr (2011a) as 16-O-p11.1/6-1; reported by the mother of the now 2-year-old girl.

I found out that I was pregnant with my third child in October 2007, a week after my daughter's first birthday. Due to my advanced maternal age (36) and the fact that my blood tests with my previous two pregnancies showed an increased risk for Down syndrome, we decided to have CVS testing done at 9 weeks of pregnancy. I had had amniocentesis done with my previous pregnancies so I was not concerned or nervous.

We met with a genetic counselor prior to the test and asked for quick results. The initial results came back free of Down syndrome and were normal for a boy. They did remind me that I would need to wait for the full results, but I put this out of my mind because of my previous two pregnancies. We started to get excited about the pregnancy, although we didn't yet share the news.

About a week later I received a call from the intern. She explained to me that my CVS came back with a mosaic marker (sSMC), an extra piece of chromosome, and that we would need to have further testing done to see if it was anything serious or not. Thus began our long and stressful journey. I spent the next 2 months nervous, worried, and upset. I cried a lot and my husband and I had a lot of discussions about our options and what we should do. Both my husband and I had to be tested to see if we had the marker. We did not. Further testing showed that the marker came from chromosome #16. Next we had to wait to see if this marker was considered to be euchromatic or heterochromatic material.

During the next month and a half I learned all that I could about sSMC. I found the information to be very confusing, and until I knew which chromosome was involved I felt overwhelmed. I asked my genetic counselor for more information and we scheduled a time to meet with her right after Christmas. During our meeting with the counselor we looked at prior cases of marker 16. She could only find about five and I think that three of those pregnancies had been terminated. There really wasn't anything we could look at to know what the future would hold for the baby. We also decided to have an amniocentesis done at this time to see if the results would be the same.

While we were waiting for the amniocentesis results I kept reading the information that the counselor had given to me. And I finally came across a web address that led me to the website on sSMC (http://www.med.uni-jena.de/fish/sSMC/00START.htm; Liehr 2011a). I was so relieved to finally find a compilation of more than five cases dealing with chromosome #16 and here was a site dedicated to sSMC!

Around the same time that I found this site, my counselor called to tell me that the results had come back as heterochromatic, so the marker should be benign. However, the lab report stated that they could not be 100% sure and they said that there was some further testing that could be done but that the lab we used did not do it. I was 17 weeks pregnant and we had just 3 weeks left to decide what we were going to do. At that time I contacted Dr. Liehr. He offered to test some of the cells and I contacted my counselor about this. She had the lab sent the material to Jena and as soon as he had results he sent them to me. They were the same as before. He reminded me that no one would give me a 100% guarantee but that the material appeared to be heterochromatic.

At 20 weeks we had an ultrasound and we decided to continue with the pregnancy. All of the ultrasounds we had had showed no problems. The baby was growing at a normal rate and was active. As the pregnancy progressed we managed to try to forget about the findings and we prepared ourselves for another little blessing.

My doctor had the neonatologist there during birth just in case, but when our son was born he was given the all clear. He was my heaviest baby, weighing 8 pounds and 2 ounces and was 19 inches long. He was beautiful. He had a head full of hair and looked a lot like his older brother. My husband and I were both very relieved, although I knew that neurological issues couldn't be ruled out. Since I was an experienced mother I thought I knew what to expect. He truly was a normal baby in every respect, although my pediatrician did say that his ears appeared to be slightly low set (they are not). He was a bit fussy in the hospital and after we had taken him home he got even fussier. He cried a lot and ate a lot, but after eating he didn't fall into a peaceful sleep. His first 10 weeks were rough – he was admitted into the hospital at 2 weeks of age due to a high fever (which I think he caught from his siblings). I was assured by the doctor that it had nothing to do with his marker. He ended up on a hypoallergenic formula, which helped his sleeping tremendously. It took him a long time to sleep through the night (12 weeks). When he was awake he had to be held or he was miserable. For a while the only place I could get him to sleep was in the swing. He didn't start taking regular naps until he was about 5 months old. This was not normal for me, but finally after those initial 12 weeks he began to change.

He turned into a loving, calm, interested, joyful baby. He has big, beautiful blue eyes (so far) and loves to smile and laugh. He has met all of his milestones. He is our best eater as he seems to love all food. He has no allergies that we are aware of at this point. He just turned a year old and although he's not yet walking, he does cruise the furniture and he loves to pull himself up to things. He is communicative, although he only says "dada" at this point. He babbles constantly and easily entertains himself. He is truly a blessing and at no point since he was born (other than when he was admitted to the hospital) have I worried that the marker chromosome will have an effect on him. My husband and I joke that he is our most "normal" child – our son didn't talk until he was 2½ and our daughter didn't walk until she was 16 months old. At his 1 year appointment we were told that he needs to gain weight as he only weighed 18 pounds. However, when you look at him he looks chubby and healthy so we are supplementing with some special food and not worrying about it.

This pregnancy and experience was a learning process for my entire family. It brought us closer together and I was relieved to go through with the pregnancy and at least be able to add to the information out there for others who find themselves with the same marker that our son has. I find myself looking around at people now and wondering if they may have a marker that they don't know about. If we had never had the CVS or amniocentesis done we wouldn't know today that our son has an sSMC. We are truly blessed with this little boy and we are grateful to Dr. Liehr and to all those who helped us throughout the process.

4.2.4 Personal Experience 4

Case listed in Liehr (2011a) as 14-O-q11.1/1-14; Poland.

We have just a short story to tell. My husband and I had been married for several years already and wanted to have children; a wish which unfortunately was not fulfilled. So we went to genetic counseling and our chromosomes were studied from peripheral blood. By that I was identified to be carrier of an sSMC. In genetic counseling we were told that due to the sSMC presence I would never be able to get pregnant, a message which was quite frustrating. As we were not satisfied with the consultation we were looking for further information in the internet and found a specialized laboratory in Jena, Germany. There we sent our blood samples as well as those of my parents. It turned out that my mother had the identical sSMC as I have and by that it was clear that the sSMC could not be the reason for our unfilled wish to have a child.

4.3 sSMC in Genetic Counseling

From the real-life reports above (see Sect. 4.2) it is clear that diagnosis of an sSMC, like that of any other chromosomal abnormality, has a profound impact on the whole family involved. Thus, the genetic counselor has a highly responsible role in this scenario. He or she has to find the balance between providing all available information to the couples/the sSMC carrier and presenting it in such a way that it really reaches the counseled persons. Most critical are the prenatal sSMC cases, as there is always the problem of limited time. However, as is clear from the fourth case report, even postnatally detected sSMC, if patients are counseled in such a way that they might, e.g., misunderstand the counselor, can have adverse consequences for generations.

In this book it is not intended to review all the ethical issues, legal problems, and quality issues (Kristoffersson 2008; Kristoffersson et al. 2010) connected with genetic counseling. However, some important points should be mentioned here in connection with sSMC in human genetic counseling.

Some national human genetics societies (GfH and BVDH 2007) recommend genetic counseling is done before each genetic analysis, and this has even been regulated by law recently, e.g., in Germany (Köhler et al. 2009). For sure, the possible detection of an sSMC cannot be a topic of this kind of general consultation; however, if a chromosomal analysis is intended, it should be mentioned to patients or persons seeking advice that aberrant karyotypes might be detected even in healthy persons or in an apparently "normal" pregnancy.

After an sSMC has been found, such a result should, if possible, be communicated in a face-to-face situation and not by e-mail, fax, or telephone. The counselor should make sure that the counseled person understands the problem and should offer the possibility to interrupt the conversation if it "all becomes too much" for the moment. Especially in prenatal cases one has to be aware of emotional reactions. As the late Prof. Uwe Claussen (Institute of Human Genetics Jena, Germany) always said to his students: "If prenatal diagnostics finds any chromosomal aberration which has an impact on the fetus' health, you as MD have to know that for the parents in this moment a sense of bereavement is felt. The expected healthy child has died! So now the parents first have to have time to 'bury' this child mentally, before they can accept the real fetus with health problems." Thus, the time after giving a chromosomal diagnosis such as the presence of an sSMC has to be followed carefully and supportively by the genetic counselor.

Certainly, the counselor has, as soon as the person(s) seeking advice is/are prepared, to provide information about all known facts on sSMC and offer additional tests. These should include parental chromosomal analysis and how and where the chromosomal origin and content can be characterized; also, if appropriate the problem of UPD should be discussed. It is also often helpful if the patients make contact with corresponding support groups (see Appendix).

Chapter 5
Small Supernumerary Marker Chromosomes Known To Be Correlated with Specific Syndromes

Classically, four clinical syndromes are included in the cytogenetically defined group of patients with an sSMC: ES (see Sect. 5.1), CES (see Sect. 5.2), PKS (see Sect. 5.3), and i18pS (see Sect. 5.4). Besides, there is the special group of patients with mosaic karyotypes 45,X/46,X,+mar who can, but need not have, TS phenotype (see Sect. 5.5). Isochromosome 8p syndrome (see Sect. 6.8), isochromosome 9p syndrome (see Sect. 6.9), isochromosome 15 syndrome (see Sect. 6.15), and isochromosome 20p syndrome (see Sect. 6.20) are sometimes mentioned as well. However, these extra chromosomes are larger than chromosome 20 of the same metaphase spread and are thus, by definition, SMC because they can normally be identified by banding cytogenetics alone. Finally, UPD-related syndromes that might coincide with sSMC presence are also discussed in the corresponding subsections in Chap. 6.

5.1 Emanuel Syndrome

ES (OMIM #609029), also known as "derivative chromosome 22 syndrome", is most often caused by a balanced translocation of chromosomes 11 and 22 in one of the parents. This translocation, cytogenetically described as t(11;22)(q23;q11.2), is the most frequently occurring recurrent non-Robertsonian constitutional translocation in humans (Zackai and Emanuel 1980). It was demonstrated that palindrome-mediated double-strand breaks in meiosis cause illegitimate recombination between subbands 11q23 and 22q11, resulting in this recurrent translocation (Kurahashi and Emanuel 2001). Also, one low copy repeat (LCR) on chromosome 22 has been shown to be involved (McDermid and Morrow 2002). There are hints that the ES-specific translocation of chromosomes 11 and 22 is predominantly formed de novo during male spermatogenesis (Kurahashi and Emanuel 2001; Kato et al. 2006). Carriers of the balanced constitutional t(11;22) translocation are phenotypically normal. However, there is a 2–6% risk of their having live-born progeny with ES (Medne et al. 2010). The latter is a result of malsegregation of the derivative chromosome 22 during meiosis (Shaikh et al. 1999). In the literature more than 300 ES cases have been reported (Liehr 2011a). The ES-causing der(22)t(11;22)(q23; q11.2) is detected in cytogenetic analysis as a centric minute-shaped sSMC (see Sect. 3.2).

5.1.1 Clinical Characteristics

Clinical manifestations are/can be microcephaly (100% of ES patients), cardiac defects (60% of ES patients), cleft palate (50% of ES patients), renal malformations (30% of ES patients), anal atresia or stenosis (20% of ES patients), hip dysplasia, other skeletal complications, craniofacial abnormalities, gastroesophageal reflux, hearing loss, refractive errors, strabismus or other ophthalmologic issues, inguinal hernia, seizures, and other manifestations. Life-threatening congenital malformations such as congenital heart defects, diaphragmatic hernia, and renal insufficiency may lead to early death during the neonatal period or even during the intrauterine period. Because medical care has improved survival chances, adult ES is reported nowadays (Medne et al. 2010).

Patients with ES have severe global developmental delay and/or severe to profound mental retardation. That means some individuals are able to use single words to communicate; most individuals can sit but not walk. For therapy, first standard (surgical) therapy of birth defects, such as cardiac and renal problems or cleft palate, has to be done. Later, ongoing physical, occupational, and speech therapies and, if appropriate, alternative communication methods to facilitate communication can be used (Medne et al. 2010).

5.1.2 Cytogenetic Characteristics

ES patients have genotypically an unbalanced chromosomal situation owing to the presence of a derivative chromosome 22 described as der(22)t(11;22)(q23;q11.2), i.e., they have a so-called complex sSMC (see Chap. 10). This leads to partial trisomy of 11q23 to 11qter and 22q11.2 to 22qter; in other words, approximately 12 Mb derived from chromosome 11 and approximately 20 Mb from chromosome 22 are additionally present in ES patients. Thus, if a centric minute-shaped sSMC derived from chromosome 22 is characterized by centromeric probes, the possibility of ES has to be considered. A parental chromosomal analysis can answer the question of whether one of the parents is a carrier of t(11;22) or not. Also, cytogenetically ES patients are rarely mosaic but rather have the sSMC in every cell.

5.1.3 Patient Report

Provided by Unique; reported by the mother of a 8-year-old girl with ES.

Lucy [*Note*: *most of the children's names have been changed in accordance with their parents' wishes*] *was born, following a relatively uneventful pregnancy, 2 weeks early, by elective caesarean section. The day prior to her birth I had been to hospital for some monitoring and the dreaded words "chromosome disorder" were mentioned. At that point all of our hopes, dreams and ambitions were on hold. We went into panic mode, so much so that I managed to crash the car on the way to buy super small sized nappies for a baby I had been told would be very little.*

Lucy was born on Friday morning, 5 April, and for 16 blissful hours we cradled our gorgeous, 4lb 13oz bundle and loved every minute. Then the medics realized there was a problem. Lucy was admitted to the special care baby unit and from there to Guy's Hospital in London. She had problems feeding and seemed to be slipping away from us. It was a year later that her pediatrician accidentally dropped into conversation how close to losing her we had been – very close indeed. But, something pulled her through – I like to think it was sheer, bloody determination on her part, the medical staff's and ours. It was probably the worst time in our lives. Watching our little girl being poked and prodded, tube fed, injected. The worst point was when for the best part of an hour we listened to her scream as staff tried to put a line in to get blood and give fluids, while we were watching by helplessly. Gradually it did get easier, she took small amounts of sugar water, then breast milk and then came the day when she began feeding normally. We had been determined that I would feed her myself and that's just what I did, against all expectations that Lucy would cooperate after 5 weeks of tubes. In between, a karyotype typical of Emanuel syndrome was determined: 47,XX,+der(22)t(11;22) (q23.3;q11.23)mat. At this point she came home and life returned to normality.

I use the word normality very deliberately. That is what our over-riding ambition for Lucy has been. As teachers within the field of special needs ourselves, we had very definite ideas about our daughter being an integral and accepted part of her local and wider community. Of course, there were loads of appointments. Lucy has hearing, visual, communication, heart and kidney problems as well as a general learning delay. We initially spent hours in waiting rooms, often giving the same information over and over again to people who had her huge pile of notes in front of them. Everyone was efficient and supportive of her needs and often pleasantly surprised at the progress she was making. Very early on, Lucy had been seen by a geneticist at Guy's who had explained that her needs would be mild to severe. A year later, he said that her needs would be mild to moderate. When we questioned this difference, he said that she had been dealt her hand – what happened now was down to environment, what we did and were doing about it. Some people said that Lucy is lucky to have two people in the know as her parents, because we can push her in directions and obtain support that other parents aren't so aware of. This may be true or it may just be that when you are dealt your hand you just have to say, "Hey look folks, this is it, let's make the best of it" and mean it.

Lucy has been fortunate in that everyone to date connected with her learning has welcomed her and the challenges she brings to their provision. Lucy went to a mainstream nursery school and had a marvelous key worker who was as determined as we were to make sure she had the same opportunities as other children at the setting. So Lucy learned to overcome tactile defensiveness when touching new objects, she learned to stand up using a frame to play with toys, she learned to use a walking frame to take those first few steps. And the day she learned to roll over everyone saw it and applauded her for it. Nursery made allowances for Lucy to be normal. Lucy was accepted by everyone at nursery: children, staff, parents and carers. So much so that 4 years on she is still greeted with genuine warmth when she meets them around town, people are interested in her progress, her achievements and in her as a person.

When choosing a school, we wanted to replicate this situation, to continue to see Lucy as the center of consideration both in terms of caring for her needs and in terms of her progress – academically and socially. After much soul searching we spoke to the head at the school that her mum taught at. When he said, "I thought you'd never come to your senses and ask" we could have whooped with joy. I knew that the school wasn't necessarily the best in the area, not getting the best results in exams, dealing with some tricky individuals and lots of social problems. But what it did have was the right ethos. There was a determination to make it work for children, children were at the heart and nothing was too difficult to make sure a child succeeded whatever the needs.

Changes had to be made before Lucy started in a normal, mainstream class with 18 other 4 year olds. We had started the process of applying for a Statement of Special Educational Needs when Lucy was in nursery and she came to school with 30 hours support from a teaching assistant. Caring for a child with complex needs is not easy for any assistant and it takes someone special and determined to do this job with one child day in and day out. But after some initial difficulties, we found a fabulous lady who became one of Lucy's greatest advocates and friends. Staff were trained in Makaton signing (a sign language), children learned all songs using Makaton so that they could communicate with Lucy and so that she was integrated wherever possible in lessons and assemblies. Physical changes were made to the buildings – ramps were put in and routes around school leveled. Lucy has been lucky in that her teachers are open to ideas and advice and when things have been challenging they've found ways to make them less so. The joys of seeing Lucy at school are immense and what a privilege to be able to see this on a day to day basis within school. Seeing assemblies and performances when Lucy joins in with everyone else and the tear-jerking Nativity plays whether she is a cow in the stable, an angel or the star are wonderful. If anything, mainstream school is absolutely right for Lucy. She has had to become normal, had to learn to do things in order to keep up with her peer group. We are absolutely sure that she has progressed at the rate she has because she hasn't wanted to be left out. And, goodness me, she isn't. At her recent birthday party 11 children from school attended and she was shepherded around an indoor play park by several little girls who pulled and pushed her up steps and ramps and even tried to lift her onto swings in order that she could join in. Isn't this what every parent wants for their child – for them to be a part, to be included? School and school life is a rehearsal for adult life. Lucy's acceptance and inclusion at this time gives her skills and relationships that will prepare her for later on.

There are other ways that we try to help Lucy to be normal. She loves water and always has. So we have swimming lessons, and again the staff of the local pool have really worked hard to ensure that she gets what everyone else would. Lucy also attended Beaver Scouts and recently was invested as a Cub Scout, attending with her brother. Lucy has done many things, from parades, to overnight camps, to fund raising bag packing sessions – and loves it. How satisfying to see her babbling away to shoppers as she throws their eggs into their carrier bags, interacting with members of the public in a way I wouldn't have guessed was possible on that day before her birth.

Of course, it hasn't all been plain sailing. It would be a lie to suggest it was. There's a lot of truth in the article "Welcome to Holland" which we urge you to read if you work with any children who are different to the norm (see http://www.our-kids.org/Archives/Holland.html). There are times when I look at Lucy (Fig. 5.1) and wonder what it would be like to have a normal 8 year old, going to ballet lessons, playing with friends in the park without an adult to help, talking, running . . .. But the big question is would I change her? No, never, not now, well, maybe the troublesome kidneys or the lack of speech or the behavioral difficulties like scratching, but overall? No. Lucy has brought so much to us. Even as I write this, she's trying to distract me, recognizing her name on the screen and babbling to me to try and find out what I am doing and why, and I feel immensely proud that she's doing that – reading over my shoulder when no one expected her to be able to do that! Lucy came to us for a reason, I don't know what that reason is yet, I don't know if I ever will, but I'm glad of it.

There are lots of people who have helped and supported us on our Lucy journey. There are many who enjoy her progress as much as we do. The health service staff are a marvel. They have been professional yet humanly supportive throughout – something that was desperately needed in the early days. Nursery and school staff, Riding for the Disabled Association and scouting volunteers who have made so much possible, and had that grim determination to make Lucy succeed. There is a commonality with all of us – determination to make Lucy succeed and be a normal, integrated member of society. Sometimes that's meant being unequal, in order to bring about equality. Sure, Lucy gets more than others, in terms of time, support and opportunities, but when it brings about that success, progress and normality,

Fig. 5.1 Lucy helping in the kitchen (copyright Unique)

then it's necessary. In Lucy's terms the journey has only really got going ... long may it continue. We look forward to being with her on the next part of it.

5.2   Cat Eye Syndrome

CES (OMIM #115470) has another three designations: Schmid–Fraccaro syndrome, chromosome 22 partial tetrasomy, and inv dup(22)(q11) syndrome. The fact that CES patients most often have a coloboma of the iris, which may make the eye look like the eye of a cat, is eponymous for this syndrome (Fig. 5.2). CES is interfamilial and intrafamilial, characterized by a high variability in clinical expressivity of symptoms. That is, only approximately 40% of CES patients present with the classic triad of anal anomalies, coloboma of the iris, and preauricular skin tags or/and pits. As even in asymptomatic persons mosaicism is possible for an sSMC, the CES-typical sSMC can principally be either de novo or derived from one of the parents (Schachenmann et al. 1965; OMIM 115470). Thus, there are many more CES-sSMC carriers in the human population than the approximately 200 reported cases (Liehr 2011a). There is evidence that in de novo cases the meiotic error predominantly happens during oogenesis (Magenis et al. 1988).

5.2.1   Clinical Characteristics

CES is characterized clinically by a combination of anal atresia with or without fistula (in approximately 80% of patients), downslanting palpebral fissures, preauricular tags and/or pits (in approximately 85% of patients), and iris coloboma (in approximately 60% of patients); additionally, the occurrence of heart malformations (approximately 60% of patients) and renal malformations (in approximately 70% of patients) and normal or near-normal mental development (Kunze 2009) have been reported. According to Turleau (2005) mild to moderate intellectual deficit is present in approximately 30% of patients. Heart and renal problems as well

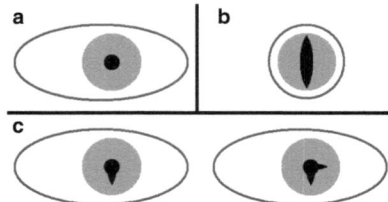

Fig. 5.2 Scheme for how the eye of a CES patient may look. *Black* pupil, *light gray* iris, *dark-gray-surrounded area* visible part of the eye. (**a**) Normal human eye. (**b**) Normal eye of a cat. (**c**) Two examples of how eyes of a CES patient may look like due to coloboma

as anal atresia can even be lethal in the early postnatal period when they are severe enough and/or are not treated. Overall, life expectancy is not significantly reduced for patients with few or mild manifestations (Turleau 2005). Only mild prenatal growth retardation occurs; however, preauricular pits, heart malformations, and renal malformations as well as sirenomelia may suggest CES. Different kinds of vision impairment can also be found, including (unilateral) microphthalmia, and also cleft palate and other malformations (OMIM #115470; Kunze 2009).

Most CES patients are borderline normal to mildly retarded, a few are normal, and some are moderately to severely retarded, although the last is rare. Behavioral problems have been reported in individual cases, but are not characteristic of the syndrome (Schinzel et al. 1981). As treatment, apart from individual support for mental development, for patients of short stature, additional growth hormone therapy might be indicated (Pierson et al. 1975).

An association between iridal coloboma and anal atresia was probably first noticed by Haab (1878), and was recognized again in 1965 (Schachenmann et al. 1965). As already mentioned, the variability of clinical features, particularly congenital malformations, is enormous (Schachenmann et al. 1965; Schinzel et al. 1981).

5.2.2 Cytogenetic Characteristics

The CES-typical sSMC is always dicentric and can be visualized by GTG banding, CBG staining, and NOR staining. Further, it can be characterized using centromeric, whole chromosome paint, or suitable centromere-near probes specific for chromosome 22 (Liehr et al. 1992). As the CES-critical regions are proximal to the DiGeorge syndrome critical region (Mears et al. 1994), commercially available probes for the DiGeorge syndrome critical region are not suited to characterize CES-typical sSMC. In CES, the sSMC can be lost during mitosis, and thus cell mosaics are observed regularly.

The CES-related sSMC evolves normally by interchromosomal U-type exchange (see Sect. 3.1). This finding, based on observation of centromeric heteromorphisms (Magenis et al. 1988; Liehr et al. 1992), is also supported by the fact that the sSMC can be asymmetric, which is more likely in interchromosomal rearrangements than in intrachromosomal rearrangements. In these cases most chromosomal regions are present in two copies on the sSMC, but some of them, in the more distal 22q region, are present only once because of unequal crossing-over events (Mears et al. 1994). Recently it was found that there are four critical regions in which the crossing over can take place, leading to different-sized variants. In the smallest type I, the sSMC are symmetric, with both breakpoints located within the proximal interval. For the larger type IIa and type IIb (and maybe type III) sSMC can either be asymmetric, with one breakpoint located in each of the two intervals, or symmetric, with both breakpoints located in the distal interval. Since the phenotype associated with the larger duplications does not appear to be more severe than that of the smaller duplication, determination of the type of CES sSMC does not have prognostic value (McTaggart et al. 1998; Bélien et al. 2008). As in ES, several LCR (McDermid and Morrow 2002) seem to be causative, so these regions are prone to a certain instability; they are located in 22q11.2 at approximate positions 16.5 Mb (type I), 18 MB (type IIa), and 20 Mb (type IIb) and for type III distal from 24 Mb in 22q12.1 (Bélien et al. 2008).

Rarely, there are CES cases with intrachromosomal tandem duplication; however, in some patients with clinically diagnosed CES, neither duplication nor CES-typical sSMC can be found (Turleau 2005).

5.2.3 Patient Reports

Report A: *See Sect. 4.2. Personal experience 2.*

Report B: *Reported by the parents of a several months old son with CES.*

A little brother or sister for our 2-year-old daughter was our great wish, and thus, it was a nice Christmas present when we learned in December last year of the pregnancy. In January the gynecologist "discovered" in routine sonography control two living organisms. As our first daughter was born with Down syndrome we opted for an amniocentesis. And "we had 2 weeks of happiness" we said, when we were contacted by the clinic 15 days after we had obtained the normal quick test result. But now we received the information that one of both twins was carrier of a marker chromosome (sSMC), and a parental cytogenetic analysis was suggested. The idea behind was that we as parents were possibly also carriers of the sSMC and then it could be considered as harmless in the fetus. Which kind of sSMC it could be and what it could mean the MDs did not discuss with us. We were asked not to be worried and to wait for the final result of human genetic analysis. That this was no option for us is obvious. We were, on the contrary, all the more concerned and started our own research on the internet. We found that sSMC could lead to a variety of clinical manifestations, according to their chromosomal origin, if they contain genetic active material, how many cells the sSMC have and if the parents are sSMC carriers or not. The end was that we envisioned a number of adverse scenarios. During the research we also got in contact with Dr. Thomas Liehr, who reacted immediately on our request, which gave us the feeling that somebody cared for our case. The same feeling we had when we contacted by telephone the Human Genetic Institute which did the amniocentesis and asked for the final results. They told us that now it was clear that the developing child had an sSMC derived from chromosome #22 and was associated with the so-called cat eye syndrome.

Again we started an internet search on this topic. Step by step it became clear that there are only few reported cases, the spectrum of possible physical abnormalities is wide and the grade of mental impairment could vary from minor to severe. We had 2 weeks with lots of worries until the clinic finally "invited" us for discussion and sonography. It has to be mentioned that in prenatal sonography some distinct physical abnormalities can be controlled "continuously", but others not, like e.g., the guiding symptom of the cat eye syndrome, the anal atresia. In first sonographic scan some abnormalities in brain and facial profile as well as growth retardation compared to the second twin were detected. But all of them were quite discrete and it should stay like that during the whole course of the pregnancy.

But what did these findings mean for us? Before we did amniocentesis it was quite clear that we were not prepared to live with another handicapped child, as already now lots of our forces were involved in the care for and support of our daughter. Could we stand that a second time with twins; possibly for a child with more severe physical abnormalities up to the risk of being dependent on care? Would our present and future family with all its members "survive" this? And what to do in our special case, with an affected and a non-affected twin fetus? Performing a termination of the pregnancy for the one and carrying the risk of losing both of them? Doing a so-called selective termination? If so, at what time of the pregnancy could it be done, to reduce the risk for the non-affected child as much as possible? And who in Germany would be willing to do that?

The head of the Human Genetic Institute doing amniocentesis contacted us and again we had the feeling not be left alone in this difficult situation. Many conversations, individually, by telephone, by mail followed. We contacted Dr. Liehr, also another specialist in human genetics, other MDs at the local and at two other clinics in Germany. Contacts to families with similar affected children were initiated by LEONA, an association for parents of children with chromosomal aberrations (see Appendix). Additionally, we utilized psychological support, as we could not imagine being able to live on having done an induced termination of the ongoing pregnancy, especially without knowing how much affected the child would be. Another option we could possibly live with would have been giving birth to the child and to give it up for adoption or for a foster home. So we collected information for that, as well.

After long weeks of consideration finally we came to a decision against the selective termination and for the affected child. We wanted to give birth and wait and see if we would be capable of giving appropriate care to the child; this decision should be made according to the extent of inborn aberrations. Based on that, the last weeks of the pregnancy were somewhat calmer, as we could not do anything but waiting.

About 2 weeks before term the twins were born and our son with sSMC was immediately examined on the new-born intensive care unit "from head to toe". He had "suffered" a little bit, as during the last weeks of the pregnancy he was located below his brother, had been the smaller and weaker, and was "crinkled" with wry neck and somewhat deformed skull. But luckily many of cat-eye-syndrome associated symptoms could be excluded in his case. What remained was at that time a small palatine cleft in the back part of the palate; all other checks were negative. A nuclear magnetic resonance imaging (NMRI) will give insight into the structure of his brain. Thus, the little man could be sent to his home 10 days after his birth, where he was awaited by his sister and his twin brother. Since then he has been fed up, which made difficult to all involved parties by his permanent spitting from nose and mouth. The present goal is to reach 5–6 kg of body weight for the *pending surgery of the palatine cleft. Unresolved are still some questions as if he will later in life develop other symptoms like, e.g., spasticity or seizures, and how his mental impairment and further development will look like. But we have learned that we have to take the things as they come and cannot beat anybody to the punch.*

Report C: Provided by the parents of a now 17-year-old son with CES.

Silas [Note: most of the children's names have been changed in accordance with their parents' wishes] was born in a hospital in Germany, where we (both parents) have made our education as medical orderly. The pregnancy was entirely normal. Silas was born spontaneously with an APGAR of 8/10/10, however, he was remarkably silent. During a pediatric routine check 3 h after birth the MD found that something was wrong in the anal region of the baby. After a short and good consulting talk Silas was transferred to another clinic together with me (his father), suggesting he just had a bowel occlusion. Arriving in the clinic I was shocked by the high-tech medicine. From that moment on we went accompanied by an emergency doctor and oxygen from one clinic to the next. I was present during examinations and was involved and explained what MDs found out. At the end of the day it was clear by sonography that Silas had a 7 cm high anal atresia with fistula to the bladder, and that an artificial anus surgery was necessary. This information was surprising for me but tolerable, as the problem was surgically treatable, and we trusted in the medical care system. I left the clinic this night to inform my wife in the women's hospital.

The real shock hit us the next day. After the surgery Silas has had several apnea. His general condition was critical, a heart defect was not excluded. That was too much for us. From now on each medical round was a dread, as each day brought new startling news. He had an ASD, was too weak to be breast fed and a cerebellum anomaly was suggested. After several days our son opened his eyes for the first time and I saw a slit in his pupils, resembling a keyhole. We gave this Information to the MD and he induced a detailed examination. Afterwards, again, we were taught another bad news: Silas would be color blind or possibly without vision at all.

At the age of 4 weeks Silas got a place for a surgery at the open heart. We had to provide our consent and had a corresponding talk. We were told that he lethality rate was at over 20% and later he should not do any sports; a mental impairment could not be excluded due to the operation either. Nonetheless, we knew that the surgery would be the best for our child. We were recommended not to see Silas for the next 3 days, as he would need rest and his body would be in an unhandsome state due to the surgery, as well. In the evening after the surgery we learned that Silas had survived but did not have an own heart frequency. He had a cardiac pacemaker, which, thank God, could be removed after 10 days. From now on the development was relatively positive. At the age of 6 weeks he was released from clinic. Shortly before his release a professor of Human Genetics gave us the complete diagnosis. He told us all physical defects were due to a single additional derivative chromosome #22, with led to the so-called cat-eye-syndrome. Thus, the "keyhole" in the eye.

The release was a pleasant event, but with the bitter after-taste of the question where the sSMC(22) came from, and if any of the two parents of us were carriers.

At home we were looked after by the blind school and by ergotherapy. Soon it became wonderfully clear that Silas was not blind. It remained to be excluded if he was color blind and what about the mental development, especially as Silas was developing very slowly concerning his motor abilities. We still remember the statement of a medical professional that the motor development problem would become more severe in future. It stayed like that until Silas got a heart pace maker. Before, with 11 months, Silas had another surgery in which the artificial anus was replaced by a normal one. Up to this moment we did not know any other affected people with whom we could swap ideas. The support we got was mainly symptomatic during in-patient stay.

Soon we had the sense that it would be not good for Silas to grow up as only child. We got two further children, which were healthy and went, with MD's consent to tropical Eastern Africa. There Silas had two further emergency surgeries due to bowel and bladder problems.

Shortly before enrolment, back in Germany, Silas was 6 years old, we were invited to a kind of medical consultation. In a more one-sided talk we were suggested to be genetically tested as parents. We were given a few hours to think about this possibility. We felt emotionally overwhelmed and would have liked to have a neutral utterance. However, we felt the responsibility for our two other children, who could theoretically be carriers of the sSMC, without showing symptoms, and so we opted for the chromosomal study. After long weeks of anxious waiting we received a letter, which we hardly dared to open. Nonetheless, the result was in favor for us, saying that we were no sSMC carriers and our healthy children would not have to deal with this problem later.

Still, we did not know where to go now to obtain further support. It would have been good if, beside genetic diagnostics, the hospital or a social service would have accompanied us. We asked us the same question again and again: what was the reason for this mutation in our genetic material. The family anamnesis was empty. Even though we were not medically laymen, we could not find any according subject literature at this time. We were told that there were only 62 reported cases with CES. However, we were not in need of scientific but emotional support. The latter we found in a believing in God and my theological studies, which I started in between.

In retrospective, we are thankful that the cytogenetic investigation took place and we do not need to have any concerns with respect to our meanwhile three healthy younger children. We are thankful to God, that Silas can now go to class 10 of a mainstream school. He can, irrespective of his second heart pacemaker, be active in sports. He is not color blind but has a poor eyesight, not completely correctable by his glasses. Also he suffers from a partial fecal incontinence, due to the weakness of sphincter muscle, which is treated by physiotherapy now. He is socially a little bit labile, has slight disorientation and difficulties to concentrate and has a ADHD (attention deficit hyperactivity disorder). Nonetheless, he is an original we would not want to miss.

Report D: Reported by the mother of a school child with CES.

After an uncomplicated pregnancy our child was born in spring some years ago in a hospital. It rarely opened the eyes and had feeding problems. Because of an icterus it was referred to children hospital at day 4 of life. There they recognized "something being wrong" and an investigation in anesthesia was undertaken. During that coloboma of iris and retina were found; only half of macula and retina was present. We were told our child would be blind, but it would be ok apart from that.

We really had to "digest" this information, as it was our first child. Certainly our friends and family were shocked, as well. However, after accepting the situation somehow, we quickly made contact to patient organization via or midwife. The latter had a cousin with two blind children and the cousin, we did not know, told us to immediately start with early intervention therapy.

In between, we received the result of the cytogenetic blood test performed in our child and both of us parents: a partial tetrasomy 22 was found, i.e., a cat-eye-syndrome (CES). And we were told for another child there would be no advanced genetic risk to have CES, as well.

According to the advise of our midwifes cousin we started with early intervention therapies, one for sight training and one for orthopedagogy. Nowadays we can support the idea to start these things so early, as they were very positive for the development of our child.

Being 3 years of age our child got a healthy sibling.

It was our idea that our child should not grow up as some kind of "special" being, thus, we sent it to the local Kindergarten. This we had to fight for, as the Kindergarten felt our child would need a special Kindergarten. And this fight continued when elementary school started. We only got support by one schools inspector. The teachers were doubtful, which we could understand. Disappointing was, that our teacher for early intervention therapy tried to convince the school teachers that it would not work. Anyway, we made it that our child could go to normal elementary school and it went good, as our child had all necessary aids and appliances to follow the lessons. In the end the school administration and the teachers felt that the 4 years having our child at the elementary school was enrichment for all.

We wanted to continue that successful way in the following mainstream school. We made contact early, and the responsible of the new school visited the lessons in primary school to understand how it works with our child. Unfortunately the head of the new school changed and the always supporting schools inspector retired. Also it was not allowed to meet the new class mates before school started, which was advised by parents being in the same situation as we were.

As a result in the new school our child has no problems with the marks; however, the acceptance of the class mates is quite low. Still, our child does not want to go to a special school. So we will keep on fighting to keep our child in the normal school, as the marks support this idea.

In our daily routine we do not recognize our child has a visual impairment and we will keep on treating it as a "normal" child.

5.3 Pallister–Killian Syndrome

In OMIM PKS has the entry #601803 and was independently reported in 1977 and 1981 (Pallister et al. 1977; Teschler-Nicola and Killian 1981). Alternatively, it is called tetrasomy 12p, hexasomy 12p, or isochromosome 12p syndrome. Overall, PKS is a syndrome with dysmorphic features involving most organs; however, it is also characterized by a tissue-limited mosaicism, i.e., most fibroblasts have karyotype 47,XN,+i(12)(p10), whereas the karyotype of lymphocytes is 46,XN (Peltomaki et al. 1987; Warburton et al. 1987; Soukup and Neidich 1990).

5.3.1 Clinical Characteristics

The clinical features of PKS include profound mental retardation (100% of cases), hypotonia (100% of cases), facial anomalies summarized as "coarse" facial appearance (100% of cases), including prominent forehead (100% of cases) with sparse anterior scalp hair (95% of cases), flat occiput, hypertelorism, short nose with anteverted nostrils, flat nasal bridge, and short neck, seizures (60% of cases), and streaks of hypopigmentation or hyperpigmentation. Sparse anterior scalp hair disappears later in life (Genevieve et al. 2003; Schinzel 1991). Absence of pericardium and focal aplasia cutis in the axillary area was also reported (Zakowski et al. 1992). It is also important to know that prenatally an omphalocele may suggest PKS (Chen 2007). According to Kunze (2009), only about 50% of children diagnosed with PKS survive the prenatal and newborn period. On average, patients can live up to two decades; the oldest patient is 45 years of age. However, most are bedridden and make no progress toward language development (Kunze 2009). Even though most PKS patients are severely affected, at least one case with mild expression is known, in which the PKS-typical sSMC was present only in 37% of the skin fibroblasts (Genevieve et al. 2003).

5.3.2 Cytogenetic Characteristics

The problem of cytogenetic diagnostics in PKS is the presence of the sSMC in a low-grade mosaic in peripheral blood cells and also in amnion and chorion. When detected, its origin can be verified by FISH using any kind of probe specific for the short arm of chromosome 12. Detection is also possible in interphase cells. For the origin of the sSMC, a nondisjunction event during maternal meiosis was suggested – possibly a U-type exchange (Cormier-Daire et al. 1997; Peltomaki et al. 1987; see Sect. 1.3). In one case (12-Wpks-159, Liehr 2011a) trisomic rescue of paternal chromosome 12 was proven to be the initial event for the formation of an isochromosome 12p.

In rare cases deviations from the usual karyotype with isochromosome 12p may lead to PKS (Liehr et al. 2008b; Vogel et al. 2009). Most interesting, in three cases with neocentric sSMC derived from 12p11.22 to 12pter (Huang et al. 2007), 12p12.3 to 12pter (Dufke et al. 2001), and 12p13.31 to 12pter (Vermeesch et al. 2005) PKS phenotypes were also observed. Thus, the PKS-critical region can be narrowed down to 12p12.31 to 12pter.

5.3.3 Patient Report

Provided by Unique; reported by the mother of a 6-year-old boy with PKS.

Our son Jude [Note: most of the children's names have been changed in accordance with their parents' wishes] was born in December 2004, but our story started earlier. He was diagnosed with PKS when I was about 25 weeks pregnant. At an ultrasound he showed a cleft lip, short long bones, and enlarged brain ventricles. His nasal bridge did not look well defined either. What a shock, not only to discover our baby might not be healthy and probably would be severely handicapped but, to find so little help and information about PKS available. Cytogenetic analysis gave a karyotype result of 47,XY,+mar[9]/46,XY[6]. How hard the next few months were imagining the worst might happen.

Jude was born quite healthy and since has developed nothing more than the colds and ear infections a typical toddler would. His body is healthy and his organs are fine for now, so we can expect many more years of loving him. Jude has profound global delays and is currently at about a 5-month developmental level. He also has hearing and vision impairment, low muscle tone, and seizures. The seizures are the hardest thing to deal with as they seem to be not fully controllable. They have much improved, however, and we are thankful to a great neurologist and all the doctors Jude sees. Jude cannot sit up, crawl, communicate or feed himself (Fig. 5.3). We're only beginning to recognize some signs of self-awareness.

Jude started school when he was 2 years and 8 months old. For 3 years, he has been working hard in a great school that he attends every day. He receives

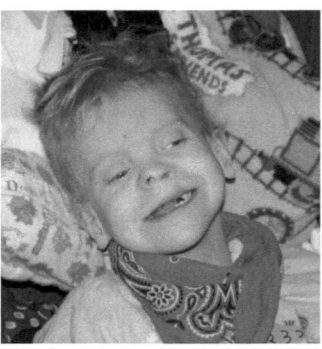

Fig. 5.3 Jude who has PKS (copyright Unique)

occupational, physical, hearing, vision and speech therapy while he's at school. When he's home, he's allowed to be himself!

Jude loves to swim and for a short time took horse-back riding sessions to help his balance and core strength. He enjoys listening to music and swinging. We've seen some nice progress lately. In December 2009 he learned to drink from a straw and in early 2010 we were seeing marked improvement in weight bearing. So much so that school is slowly introducing a gait trainer. I know Jude has a long way to go but God has blessed us with progress!

Jude has a big brother and three big sisters and they all love him to pieces. There were ups and downs early on about having a special needs sibling, but they now understand that he's still their brother and as deserving of love, care and opportunities as every other child God creates.

Five years later, we have a little boy who, yes, is very handicapped. But we also have a son who makes us smile when he smiles. He has a wonderful laugh and has a will of his own despite not being able to talk or do much, and whom we love very much. He's not perfect, but neither are we!

Raising awareness and providing hope about PKS has become a passion of mine. I'm a founding member of our non-profit organization PKS Kids (www.pkskids.com) and work hard to create tolerance and acceptance of Jude in our community as well as recognition of PKS globally. I've given several talks locally and believe I'm making a difference and giving the hope and help other PKS families need and deserve.

Jude deserves to be loved and accepted as much as any other child. I wouldn't change one bit about him. He has stolen my heart just the way he is.

5.4 Isochromosome 18p Syndrome

Over 200 cases of i18pS have been reported in the literature (Schinzel 2001; Kunze 2009; Liehr 2011a). The Chromosome 18 Registry has encountered over 300 affected individuals, and suggests an incidence of 1 in 140,000 live births (http://www.tetrasomy18p.com/). Nonetheless, the syndrome still has no entry in OMIM. It was not recognized before the mid-1980s as a syndrome, even though an isochromosome 18p was the third sSMC reported in the literature (Froland et al. 1963; Rivera et al. 1984).

5.4.1 Clinical Characteristics

Prenatal and postnatal growth retardation, microcephaly, moderate to severe mental retardation, asymmetric facial appearance, urogenital malformations, simian crease plus clinodactyly and/or camptodactyly may be observed (approximately 60% of cases according to Sebold et al. 2010). Also typical is a muscular hypotonia in newborns, changing to hypertonia of the lower extremities connected with ataxic gait. Recently, Sebold et al. (2010) reviewed clinical signs of 107 individuals with i18pS as follows: developmental delay/mental retardation (100%), abnormal muscle tone (73%), neonatal complications such as feeding problems (64%), brain magnetic resonance imaging variants (63%), microcephaly (53%), strabismus (45%), scoliosis/kyphosis (37%), recurrent otis media (35%), history of constipation (32%), growth retardation (30%), cardiac defects (24%), seizures (21%), history of gastroesophageal reflux (14%), hearing loss (12%), early death (4%), and in males cryptorchidism (39%). Children with i18pS should have periodic ophthalmologic and audiologic evaluation, cardiology evaluation, renal ultrasonography, orthopedic evaluation for management of foot abnormalities, monitoring for scoliosis and kyphosis, neurological evaluation for seizures, and referral for developmental service and therapy (see Sebold et al. 2010).

However, the phenotype can be variable; thus, rarely an isochromosome 18p may be passed through the generations; one patient with germ cell mosaicism has been reported (Schinzel 2001; Kunze 2009). Patients without any clinical symptoms due to mosaicism have also been seen (cases mother of 18-Wi-41 and 18-Wi-158, Liehr 2011a).

5.4.2   Cytogenetic Characteristics

An additional isochromosome 18p is present in some of or all cells of affected persons. sSMC(18) most often forms by a maternal nondisjunction error during meiosis II, rarely during paternal gametogenesis. As already mentioned, i18pS can rarely be inherited from one of the parents (Kunze 2009).

An i(18)(pter->q11::q11->pter) can also accompany other chromosomal rearrangements. Thus, instead of the normal partial tetrasomy 18p induced by the sSMC(18), trisomy 18p can also be the result (e.g., cases 18-Wi-43 and 18-Wi-45, Liehr 2011a).

The critical region for i18pS might be in 18p11.21 to 18q10, as the patient in case 18-W-p11.21/2-1 (Liehr 2011a) had a partial tetrasomy of this region only and symptoms such as those reported for i18pS patients.

5.4.3   Patient Report

Provided by Unique; reported by the mother of a 9-year-old boy with i18pS.

Laurence [Note: most of the children's names have been changed in accordance with their parents' wishes] was born after a trouble free pregnancy 4 weeks before my 36th birthday, a much wanted first and, so far, only child. After a 37-hour labor requiring all methods of intervention, he was born by Caesarean section. At 6 weeks, he was admitted back to hospital as a "failure to thrive" baby and for the next 10 months endured every possible test designed to discover why he was not growing above the 10th percentile, was not hitting development milestones and why his reflux was so very bad. Why, too, were we admitted to hospital so often for serious gastric infections? At 4 months, scoliosis was diagnosed, and odd, short metacarpals on each thumb. A ventricular septal defect was also diagnosed but mercifully closed itself within the year. Finally, just before Laurence's first Christmas, a test finally came back with some news. "He has Tetrasomy 18p" (47,XY,+i(18)(p10)) and, rather more firmly, "don't look at too much on the internet, speak to us first." I did of course look at too much on the internet and certainly shouldn't have. These were the key messages I remember from the emotional haze of the next few months. The ever kind and patient consultant clinical geneticist at the United Kingdom's leading children's hospital gave us my most cherished and sensible answer to my many manic enquiries: "He will just be himself". And for the last 8 years she certainly has not been wrong.

As we pass Laurence's eighth birthday, he is above the 50th percentile for weight and height but not too far to be a concern (Fig. 5.4). Thanks to medication with omeprazole his reflux is broadly under control. For the last 2 years he has worn a full spinal brace to keep his scoliosis in check. He obeyed the little literature there is on his condition and walked within his second year (he squeaked in just a month before his third birthday) and speech started to come in his third year. He is very nearly 100% dry in the day and in the last year has made huge strides with continence despite his regular and frankly unspeakable bowel movements. (The medical profession has explored all avenues and has just concluded for the second time that "That's just how he is").

As Laurence approached school age we had to be very strong and determined as the English county where we live decided he was to be allocated a school place in a local peer group school (non-specialist) with part time support. He had enjoyed full time nursery care since birth but we knew he was simply not suited to anything but specialist provision in our "Outstanding" local Special School. He did not speak and was not toilet-trained. We dug our heels in, called in support and finally won the fight. For the last 4 years Laurence has loved every day at his Special School and grown and developed enormously. By incredible coincidence a 3-year-old girl

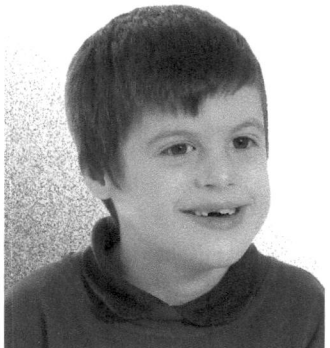

Fig. 5.4 Laurence who has isochromosome 18p syndrome (copyright Unique)

with Tetrasomy 18p has also just started at the school. We already knew that he had a local chromosome twin via the good offices of Unique and had met her as a baby when her parents contacted the organization. It is fascinating to see how many issues both children share as well as how the physical issues differ in each case. We have met two other younger Tetrasomy 18p children over the years and although the facial features and physical characteristics and movements of all the children are strangely familiar, each has a different physical issue: for example, Laurence alone of the four children we know has scoliosis.

Laurence is a big character with an incredible sense of humor and his language skills grow every day. He speaks in sentences and although some sounds are still tricky he will gamely attempt any new word he hears with often misplaced confidence and brio – accurate consonants are not his greatest skill. His balance is still weak and he is unable to balance on one leg without help despite being an expert on bouncing on beds, climbing trees and running – with or without his brace. He spends short periods in full leg splints every day at school to improve his leg and foot muscle tone and flexibility.

Laurence is very sociable and constantly craves company and stimulation. His concentration time is still short but improving and he has to be encouraged to keep focused on any activity at home or school, unless it is the TV or computer, of course. Somehow they can hold his attention for hours. He has always had a very acute visual sense. Before Laurence could speak it was clear that he recognized the most minute logos or visual clues on objects and associated them with pleasurable activities. For example the tiny logo on a DVD which represented the same brand as his favorite show or the golden arches of a burger restaurant on a far horizon.

Laurence is left-handed with short thumb metacarpals, so writing is challenging and still at an early stage. He is now, however, rather good at drawing faces and writing the first few letters of his name. He adores books and always has done and we are delighted to see some key three-letter words are now being genuinely read while he guesses or remembers the other words in his reading books – just like we all did at first!

Laurence also loves interactive role play and, as long-ago ex drama students, his parents are all too happy to indulge him with our full range of accents and voices suitable for monsters, ghosts and evil witches. Sadly he has not inherited our show-off natures and has refused to take part in any school show performance since his first days at nursery. This is not helped by the constant refrain from teachers that he was "Absolutely word perfect and fine at the dress rehearsal".

We can now genuinely look forward to the future with great hopes for Laurence's improving quality of life and the chance to help him optimize his life skills with the support of professionals at school and in the medical and caring professions. When he was diagnosed we had no guarantees that he would even walk or talk. That now feels so far from where we all are today. Our aspirations know no bounds but we know we must be content to allow him to enjoy every day as he does now and equip him as best we can for his own independent life in whatever way he wishes to live it.

5.5 Turner Syndrome

TS is present in approximately 1:2.000 to 1:4.000 newborns with female phenotype. Cytogenetically, the syndrome is most often (approximately 60% of cases) characterized by pure monosomy of the gonosomes (45,X) and thus TS is one of the most common types of aneuploidy among humans. However, in approximately 40% of TS cases mosaicism is observed in peripheral blood, with a 45,X karyotype accompanied by one or more other cell lines with a complete or a structurally abnormal X or Y chromosome (Oliveira et al. 2009; Davenport 2010). One cytogenetic subgroup of TS patients is individuals with mosaic 45,X/46,X,+mar karyotypes leading to female or male phenotypes (Liehr et al. 2007b).

5.5.1 Clinical Characteristics

TS is, as all the syndromes mentioned in this chapter, associated with a spectrum of potential abnormalities. Four areas of major clinical concern in TS were mentioned by Oliveira et al. (2009): growth failure, cardiovascular disease, learning disabilities, and gonadal failure. Growth failure may be treated by appropriate hormone therapy, allowing nowadays for rapid normalization of height. To avoid cardiac problems, magnetic resonance imaging should be performed at diagnosis and repeated every 5–10 years. Also, Oliveira et al. (2009) recommend treating hypertension aggressively. Nonverbal learning disabilities marked by deficits in visual–spatial organizational skills, complex psychomotor skills, and social skills are common in TS. That is, neuropsychological testing should be done during childhood and corresponding support and appropriate therapy and an appropriate school should be chosen. A health care checklist as provided by Oliveira et al. (2009) might also be helpful for other problems such as strabismus, hearing loss, and autoimmune thyroid disease.

For gonadal dysgenesis, hormonal replacement therapy is indicated at a normal pubertal age and should be continued until the age of 50. Transdermally administered estradiol provides the most physiological replacement (Oliveira et al. 2009). For TS patients with dysgenetic gonads, the presence of Y chromosome material detected during cytogenetic analysis (see Sect. 5.5.2) indicates an increasing risk of gonadal tumors, especially gonadoblastoma, estimated to be approximately 30%. Gonadoblastoma is a benign tumor, but it can undergo transformation into invasive dysgerminoma in 60% of cases, and also into other, malignant forms of germ cell tumors.

5.5.2 Cytogenetic Characteristics

Karyotype 45,X/46,X,+mar can be detected prenatally or postnatally. In postnatal cases, the above-mentioned exclusion of Y-chromosomal origin of the sSMC is most important to minimize the risk of gonadoblastoma (see Sect. 5.5.1).

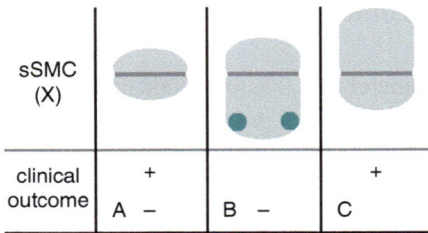

color code: euchr. X; cen X; XIST

Fig. 5.5 If a TS-related sSMC is derived from the X chromosome [sSMC(X)] it can consist of different chromosomal subregions and lead to different clinical outcomes, especially concerning mental development. (**a**) A "small sSMC(X)" with little euchromatin may or may not influence the TS phenotype. (**b**) If the XIST gene is present on a "larger sSMC(X)," this normally leads to genetic inactivation of the sSMC and thus the clinical phenotype is that of a "normal" TS. (**c**) If the XIST gene is absent from a "larger sSMC(X)," this leads most often to a more severe TS phenotype, including mental problems

Prenatally, in the case of a 45,X/46,X,+der(X) karyotype it is important to test for the ability of the derivative X chromosome to be inactivated (Liehr et al. 2007b), i.e., for the presence of the XIST (X-inactivation center) gene (Agrelo and Wutz 2010). The X chromosome is present twice in normal females. One of both X chromosomes is inactivated by expression and attachment of the XIST-RNA, except for some so-called pseudoautosomal regions. An sSMC derived from the X chromosome can only be inactivated if the XIST region is preserved on it. As depicted in Fig. 5.5, it is important to distinguish larger and smaller sSMC with and without the XIST gene, as they may lead to different clinical outcomes especially with respect to the presence or absence of mental retardation (Liehr et al. 2007b).

5.5.3   Patient Report

Provided by Unique; reported by the mother of a 12-year-old boy with TS karyotype.

Ross [Note: most of the children's names have been changed in accordance with their parents' wishes] was born in July 1997 with no obvious signs of disability. He was taken into hospital after 3 weeks as he was not thriving and at that time was undiagnosed, although he did receive physiotherapy over the next 2 years for hypotonia. At the time he was classified with a developmental delay, the cause not being known.

At the age of 16 months, he was eventually diagnosed with a rare chromosome abnormality which affected his male Y chromosome (karyotype: 45,X[2]/47,X, idic (Y)(p11.3)x2[39]/46,X,psu idic(Y)(p11.3)[4]). We had no idea how this would affect him and the only knowledge we had was that children with a similar chromosome abnormality were likely to be tall and have behavioral problems. Ross was neither tall nor, at diagnosis, did he have any behavioral issues. He was a very happy, contented baby. Unfortunately Ross's behavior altered at the birth of our daughter Sophie. He was 18 months old and started to head bang constantly. He became more unpredictable with his behavior as he got older and started more self injurious behaviors, biting his hand and slapping his hand against his head.

He received portage (early intervention) from the age of 12 months until he was accepted at a special needs nursery at the age of 2. By 3 years he was full time and benefited greatly from the structure.

His speech was limited, using odd words, and Signalong (a sign-supporting system based on British Sign Language) was used in school to help reinforce speech and language communication. He started to develop his language skills at the age of 4 and now, at 12 (Fig. 5.6), is very verbal and is able to communicate his needs well, although he has tight facial muscles which makes his speech a little difficult to understand at times.

He has learnt to swim after persevering with his swimming lessons for over 2 years and this has helped enormously with his hypotonia. He has also learnt to ride a bike without stabilizers and this has given him hours of enjoyment. He loves noisy objects like lawnmowers, strimmers and hedge trimmers and will happily stand and watch anyone who is mowing the lawn or cutting hedges. He also likes the vacuum cleaner and will spend a lot of time switching it on and off.

He lacks concentration and finds it difficult to focus on anything within the house, although he will play with play dough and his toy diggers, trains and police cars, mostly with adult support, as he is an attention-seeking child who requires a tremendous amount of 1:1 input.

In the last 3 years, Ross has been diagnosed with autism. This causes him to be very anxious and controlling. He is calmer when he knows what is happening

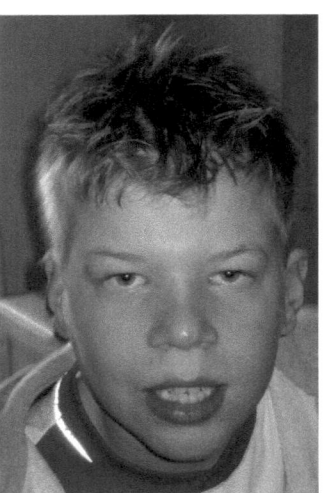

Fig. 5.6 Ross who has a TS-karyotype-related sSMC (copyright Unique)

throughout the day, although he will constantly ask about the length of time he will be doing an activity and finds it difficult when it comes to an end.

Despite his complex needs, he has developed quite a character and can often share a joke. At school he is progressing slowly and is academically at a pre-school level. He can now say the letters of his name, although he finds it difficult to write beyond the first letter, due to poor fine motor skills, and needs support to achieve this task.

He sleeps well at night, enjoys his food although he has a nut allergy, and was toilet trained by the age of 6.

At the age of 8 Ross was started on a course of medication to see if it could help lower his anxieties and therefore reduce his behaviors. He was given atomoxetine, which we persevered with for 6 months but realized that we could not see a great deal of change in his behavior. He was then tried on a course of methylphenidate but unfortunately it was a double-edged sword as it brought about seizures. He was then tried on a course of dexamphetamine but his seizures still continued and medication was stopped at this point until a diagnosis of epilepsy was given, which meant that Ross is now on antiepileptic drugs to control his seizures. Once his seizures had been controlled for over a year he was tried on a course of aripiprazole but no change occurred and he is finally now on risperidone, but a very small dose. We have never wanted to use this medication route but it was our last hope in possibly being able to manage his behavior better. Unfortunately, due to Ross's complex needs we feel that medication has not been the answer and are still trying to pursue the route of psychological intervention, but this is impossible to access locally.

Now at the age of 12, Ross's behaviors are very challenging and he displays this by biting, kicking, hitting, spitting, scratching and is also very verbally abusive. His behaviors are unpredictable and therefore he can become aggressive without warning. Ross has always been at the forefront of our lives. His needs have always had to come first with everything that we do. We have always included him in our holidays abroad, although it has been very stressful for the family as a whole, due to his continuing aggressive behaviors.

We decided last year to fly to Florida with Ross as he is passionate about theme parks and I knew that if we could get through the long haul flight, he would absolutely love the experience. It was the best decision we could have made at that time for Ross and his younger sister and brother and it is something that Ross still talks about even now.

Unfortunately at the end of last year, Ross's behaviors became too difficult to manage and he was taken into care at his respite home, where he remains at present. We are in the process of seeking a residential placement for him which has been a very difficult decision to make, but we are hopeful that he will receive the level of support he needs along with continuity of care beyond the school day.

Chapter 6
Centric Small Supernumerary Marker Chromosomes

The group of centric sSMC with different chromosomal origin constitutes about 34% of the reported cases of patients with the karyotype 47,XN,+mar (see Sect. 1.3). This group contains the most heterogeneous group of sSMC patients. Clinically healthy to severely affected individuals have been reported. In future it may be possible for more distinct syndromes such as those listed in Chap. 5 to be characterized among them.

In this chapter centric sSMC are discussed according to their chromosomal origin. For each chromosome, present knowledge on possibly noncritical regions is summarized. It is well known that there is no simple correlation, such as "sSMC without expressed genes on them (i.e., heterochromatic ones) always lead to a normal phenotype, whereas sSMC inducing partial trisomy (euchromatic sSMC) always induce clinical signs in their carriers." Rather there are genomic regions which can be present in three or more copies without causing any obvious harm, whereas other regions are apparently dose-sensitive (see Sect. 1.4.1). Here the minimum size of the non-dose-sensitive pericentric regions is given for each of the 24 human chromosomes in cytobands and according to molecular mapping. As more sSMC cases have been characterized by (molecular) cytogenetics than by molecular genetics, the regions can be larger according to cytogenetics than those confirmed by scarcer molecular data.

Nonetheless it is only possible to distinguish between "sSMC carriers with clinical abnormalities" and "sSMC carriers without clinical abnormalities." Among the latter group are included persons who are registered because of problems conceiving. As more than 50% of this group have an inherited sSMC, the sSMC is not necessarily be the reason for the fertility problems (see Sect. 1.2.2).

The attempt to draw a genotype–phenotype correlation based on the size of the imbalance induced by the sSMC is complicated by three more points.

1. Mosaicism may influence the phenotypic impact of a gene-dose effect. sSMC leading to even severe clinical problems when present in (almost) all cells of the individual may have little or no effect if present in a mosaic with normal cells. Further, duplication of an sSMC can lead to clinical problems, whereas

T. Liehr, *Small Supernumerary Marker Chromosomes (sSMC)*,
DOI 10.1007/978-3-642-20766-2_6, © Springer-Verlag Berlin Heidelberg 2012

a nonduplicated ssMC had no phenotypic impact in one of the parents (see Sect. 1.4.3).

2. UPD can be present as a side effect of ssMC formation and can lead to specific syndromes, even though the ssMC itself contains no genetically relevant material (see Sect. 1.4.2).

3. Finally, the ssMC may not be responsible for the clinical problems of an ssMC carrier, as the observed symptoms may be due to another, possibly not yet detected or studied, disease causing alteration in any gene. Such cases may lead to false-positive genotype–phenotype correlations if the ssMC is euchromatic, or to unclear results if it is heterochromatic (Nelle et al. 2010; see Sect. 9.3).

As trisomy and tetrasomy of whole chromosome arms can be present in a subset of the human chromosomes because of an ssMC the viability and/or clinical consequences of such an imbalance are also discussed for each chromosome. Finally, to highlight the variability of clinical outcomes in centric ssMC, reports provided by parents of affected children are included, like in Chap. 5. Similar imbalances such as those induced by ssMC can also be due to intrachromosomal duplications; thus, some reports are also provided by parents of a child with 46 chromosomes but also an imbalance which could possibly be induced by an ssMC (see Sect. 6.10.5).

6.1 Chromosome 1

6.1.1 Potentially Non-dose-sensitive Pericentric Region

As far as is currently known, ssMC derived from chromosome 1 lead to clinical symptoms if additional pericentric euchromatin is present outside cytobands 1p12 to 1q12 and outside the region 120.9–142.4 Mb (Fig. 6.1). The distal borders were defined carefully according to cases 01-O-p12/1-1, 01-O-p11.2/1-1, 01-O-p11.1/3-1 to 01-O-p11.1/3-5, and 01-O-p10/1-1 listed in Liehr (2011a). As outlined below, the margins are still the subject of research.

As well as clinically normal ssMC(1) patients with partial trisomy 1p12 to 1q12, there are several reports of patients with an adverse outcome (Liehr 2011a). However, as the molecular breakpoints of those cases were not characterized in detail, regions distal to 120.9 to 142.4 Mb may be involved and/or the patients may have other molecular defects such as iUPD of chromosome 1 (see Sect. 6.1.4). In the patients in cases 01-W-p11.2/1-1, 01-W-p12/1-1, and 01-W-p11.1/1-2 (Liehr 2011a), who are trisomic for the aforementioned potentially benign region, clinical signs were reported; however, in all three cases there was perinatal severe respiratory distress, hypertonic crisis with cyanosis, or cytomegalovirus infection, which also have to be considered here. Even more puzzling are cases 01-O-p11.1/2-1 and 01-O-IMB-q11/1-1 (Liehr 2011a), in which partial

Fig. 6.1 Right of the ideogram of chromosome 1 drawn according to Kosyakova et al. (2009) the possibly non-dose-sensitive pericentric region is *highlighted*. The corresponding cytobands according to the International System for Human Cytogenetic Nomenclature (ISCN) (*black bracket*) and the molecular mapping data (in million base pairs, Mb) of the region are given (*light-gray square within black bracket*). The small part marked by a *gray half-bracket* highlights a region which may be unproblematic if sSMC(1) is present in a mosaic

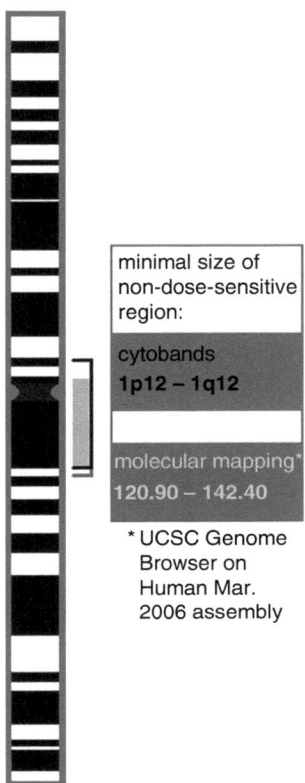

minimal size of non-dose-sensitive region:

cytobands
1p12 – 1q12

molecular mapping*
120.90 – 142.40

* UCSC Genome Browser on Human Mar. 2006 assembly

trisomy of band 1q21.1 and partial trisomy of band 1q22 respectively were reported in clinically normal patients, which is in contradiction to other comparable sSMC cases of patients with severe clinical signs: 01-W-p11.1/2-1 to 01-W-p11.1/2-4, 01-W-p11.1/3-1, 01-W-p10/1-2, and 01-W-p10/2-1 (Liehr 2011a).

6.1.2 Clinical Signs

In cases where sSMC(1) is larger than the noncritical region shown in Fig. 6.1, the most frequently observable clinical signs as listed in Table 6.1 are facial dysmorphism and mental retardation. Growth retardation might be more frequent in sSMC(1) derived from the long arm, whereas clinodactyly is seen more often in sSMC(1) derived from the short arm. However, overall the clinical symptoms are rather nonspecific.

Symptoms	1p (%)	1q (%)
Autism	22	–
Clinodactyly	33	11
Dysmorphic face	100	100
Growth retardation	22	55
Heart defect	33	33
Hypotonia	33	–
Mental retardation	78	78
Microcephaly	22	33
Seizures	11	–

Table 6.1 Symptoms regularly observed in sSMC(1) cases with clinical symptoms (alphabetic order)

Cases distinguished by centromere-near partial trisomy of the short arm and the long arm according to nine informative cases for each, collected in Liehr (2011a)

6.1.2.1   Trisomy and Tetrasomy 1p or 1q

Partial trisomies or tetrasomies of the entire short arm or the entire long arm of chromosome 1 are not viable.

6.1.3   Mosaicism

As highlighted in Fig. 6.1 (gray half-bracket), mosaicism may have an influence on the clinical outcome in cases of sSMC derived from chromosome 1. The partial trisomy of band 1q21.1 leading to no clinical signs in case 01-O-p11.1/2-1 (Liehr 2011a) seems to be because the sSMC was only present in 25% of the peripheral blood cells.

6.1.4   Uniparental Disomy

Only iUPD has to be considered as possibly disease causing in sSMC(1) cases, as no imprinted genes are known for chromosome 1. Whereas only one sSMC(1) case with maternal mixed hUPD/iUPD 1 (case 01-W-p21.1/1-1; Liehr 2011a) has been reported, more than 30 other UPD cases can be found in the literature (Liehr 2011c). Most of them were iUPD 1 and led to the activation of a recessive point mutation, and were disease-causative.

6.1.5   Case Report

Provided by Unique; reported by the mother of a 10-year-old girl with karyotype 47,XX,+r(1)(p1?q2?1)(D12S+)[22]/46,XX[28].

Sophie [*Note*: *most of the children's names have been changed in accordance with their parents' wishes*] *was born by elective Caesarean section – I had had a section with our first daughter so I was given the choice second time around. During the pregnancy there was very little movement and kicking, unlike my first pregnancy.*

Sophie was the quietest baby on the ward and because of this was popular with the nurses who did not mind looking after her for a few nights while I got some sleep! Sophie slept a lot as a baby, having afternoon naps for two, sometimes for three hours.

At the age of 3 months Sophie attended a private nursery for a few days a week while I went back to work part-time. She did not start walking until she was 19 months old. Initially she was terrified to walk unaided but we just thought it was because she was a very timid baby. At this point she was unable to crawl, stand up or sit down unaided. When she did finally learn to walk it was with a wide gait and her legs were quite stiff.

Around this time Sophie had a routine check-up at home from a health visitor. She noted that Sophie was unable to do the usual tasks expected for her age, such as stacking bricks. She also noted that her hand grip was poor and picked up on her manner of walking. We were very upset during this visit as we never suspected anything was wrong at this point. Looking back, this seems strange as we already had a 4½-year-old daughter but we just put a lot of things down to Sophie's lack of confidence.

The health visitor referred us to our general pediatrician who then referred us to a consultant at our local hospital. Around this time the words "development delay" were used when referring to Sophie's problems. I honestly thought at this point that she might still "catch up". Sophie then received weekly visits from a speech therapist and a physiotherapist. The physiotherapists taught her how to crawl, sit down and stand up, which was of great benefit.

When Sophie was around 2 years old I collected her from nursery one day to be told that she had food "stored" in her mouth from lunchtime – this was late afternoon. Sophie would not eat, drink, speak or swallow, so the next day I took her to the general pediatrician who diagnosed a sore throat although he had difficulty in getting her mouth open to examine her. She was prescribed antibiotics which were a real struggle to administer.

After a few days of this happening I took Sophie to the casualty department, who X-rayed her throat to check whether there was something lodged there. The X-ray came back clear and eventually a doctor diagnosed oral thrush and I was given a gel to squirt into her mouth. However, she had not eaten or drunk anything in several days and the doctor was concerned. He asked me to ring him the next day to let him know how she was. As it happens, he rang me next morning and because nothing had changed admitted Sophie to hospital that day.

Sophie was weak and dehydrated and put onto a drip. She awoke during the night that first night in hospital and let me feed her some snacks and a drink of water for about 15 min before "clamping up" again, still with food in her mouth. Sophie carried on periodically "clamping up" over the next few days in hospital but it was decided she would be best at home even though the thrush was not completely clear. I was to carry on administering the gel at home for a few more days.

The "clamping up" continued and after ringing the hospital and being told that the thrush should be clear by now, in desperation I took Sophie to the children's ward. Luckily her consultant was on duty and agreed to see us immediately. She was as puzzled as we were and suggested they carry out some chromosome tests. She had wanted to wait and see how things progressed with Sophie's development but under the circumstances she decided that the time was right.

The "clamping up" carried on for about 3½ weeks, resulting in myself taking time off work. I realized that Sophie would sometimes snap out of these "episodes" as the consultant called them after sleeping, so I managed to keep her fed and hydrated even though she was sometimes going for full days without food and water. The awful thing with these "episodes" was that Sophie would actually put food to her mouth but just could not seem to open her mouth to eat it, resulting in much frustration for her. The consultant described it as Sophie wanting to eat and drink but something in her brain was preventing her from doing so. The conclusion was that this was the brain's defences reacting to Sophie having a sore throat and thrush and also as a result of the trauma of the medication being administered – thus the "clamping up".

Eventually the "episodes" stopped and we got our diagnosis of supernumerary ring chromosome 1. The consultant did not have many answers to our questions as she had very little information on the condition. She also explained that each child could be quite different depending on the size of the extra chromosome and other factors.

A few years later, Sophie had a temperature so I gave her some paracetamol. She immediately refused to swallow it and this sparked off another 3 weeks of "episodes". We managed to prevent her from dehydration and hospitalization by feeding her through the night if she woke and came out of an "episode". This did not always happen – she sometimes woke up still having an "episode" which was particularly distressing for all of us.

At this time I was having a stressful time at work – this, together with what was happening with Sophie resulted in me going off sick from work – something I rarely did. I never returned to my job after working there for 13 years. The lack of compassion from my employers was very upsetting.

When Sophie was around 4, we were assigned a portage worker who visited her weekly at home. She suggested that Sophie should attend a Special Needs nursery for four mornings a week to prepare her for school. We agreed and she was transported there and back while still attending her private nursery 1 day a week. Sophie really benefited from the morning sessions.

Around this time the procedure was commenced to obtain a Certificate of Special Educational Needs in preparation for Sophie starting school. It was decided that the best choice for Sophie was to attend a Special Needs school. Sophie has thrived at the school – the classes are very small and she is so happy there.

Sophie is now 10 years old. She is a healthy, happy child. She loves music and dancing – not surprising, considering her father is a musician and her sister who is

Fig. 6.2 Sophie at the age of
10 years (copyright Unique)

13 years old excels in dance! She can pick out instruments from a piece of music and absolutely loves her "real" violin which she asked for at Christmas even though she cannot play any notes (Fig. 6.2).

Sophie has global learning difficulties and cannot read or write, although she can write one or two letters. At present she is working towards Level 1 of the English National Curriculum in all areas. Sophie can get very anxious in certain situations. For instance, she does not like sudden loud noises, especially from other children.

Sophie does not truly interact with her same-age peers. She prefers the company of her sister's friends and adults. Sophie communicates using echolalic speech and questions and can ask the same questions all day long – no matter how many times they are answered! Her communication needs are atypical, exceptional, and entrenched. Given the opportunity Sophie would watch music DVDs at home all day long. Sophie is not capable of meaningful play. She needs help with personal care – she was toilet trained fully at around 5 years old but still wears a nappy for bed. She experiences difficulty with unscrewing bottle tops and turning on taps. She needs help with dressing and undressing. Her behavior is largely passive but she can occasionally present challenging behavior such as shouting, hitting and running away.

Sophie attends a trampoline club after school which she loves. She has recently been on a school trip to London for 5 days, which she loved. We had no idea what she would be like because this was the first time she had ever been away from home, so that was a big adventure for all of us!

Sophie loves trains and swings and has an obsession with snakes, although if she sees one on the television or in real life, she hides her face! She can talk about snakes all day. Frequent questions are:, What color snakes do we like? Do they live in tanks or cages? and, Do they slither?!

Sophie is truly "unique". She can be exasperating at times but she is so cute most of the time. She can be really nice one minute then nasty the next! Either way she is our daughter and is so much loved by us and her sister. We take one day at a time – for us this is the best way of dealing with life with Sophie.

6.1.6   Chromosomes 1/5/19

The chromosomal origin of sSMC is mostly characterized by applying centromere-specific probes, which are well suited for molecular cytogenetic analysis (see Sect. 4.1.2). All human chromosomes have several so-called satellite DNA sequences in their centromeric regions; some are present on every human chromosome, others are DNA sequences specific to only one centromeric region. For example, there is one sequence that is only present in the centromeric region of chromosome 1 (D1Z5). However, another sequence (D1/5/19Z1) present on chromosome 1 is also present in the centromeric regions of chromosomes 5 and 19. Because of this biological fact, there are cases of sSMC reported in the literature that were only stainable by the probe D1/5/19Z1, specific for the centromeres of chromosomes 1, 5, and 19. Either because they were too small and other regions were lacking on the sSMC, or because the corresponding laboratory had no other probes available for the three chromosomes in question, the sSMC were reported to be derived from chromosomes 1, 5, or 19. Presently, slightly more than ten such cases are known (Liehr 2011a). For obvious reasons, for those cases no genotype–phenotype correlations can be given.

6.2   Chromosome 2

6.2.1   Potentially Non-dose-sensitive Pericentric Region

The size of the pericentric, most likely noncritical region of chromosome 2 spans a small part of cytoband 2p11.2 and (almost) all of cytoband 2q11.2. According to molecular mapping, positions 89.60–101.58 Mb are non-dose-sensitive (Fig. 6.3). Not only were partial trisomies of 2p11.2 or 2q11.2 inert, two cases with partial tetrasomies have been reported: 02-O-p11.2/1-1, 02-O-p11.1/4-1, and 02-O-p11.1/6-1 (Liehr 2011a).

Cases 02-W-p11.1/1-2 to 02-W-p11.1/2-3 and 02-W-p11.1/3-1 (Liehr 2011a) involved the aforementioned potentially non-dose-sensitive pericentric region of chromosome 2. In these cases the sSMC carriers have had clinical problems; however, the corresponding clinical descriptions lack data about the neonatal period, and/or overall the clinical detail provided is sparse.

6.2.2   Clinical Signs

The clinical signs correlated with sSMC(2) larger than the potentially noncritical regions are rather nonspecific; mental retardation and dysmorphic signs are reported most frequently (Table 6.2). Strikingly, microcephaly is exclusively seen

Fig. 6.3 Right of the ideogram of chromosome 2 drawn according to Kosyakova et al. (2009) the possibly non-dose-sensitive pericentric region is *highlighted*. The corresponding cytobands according to ISCN (*black bracket*) and the molecular mapping data (Mb) of the region are given (*light-gray square within black bracket*)

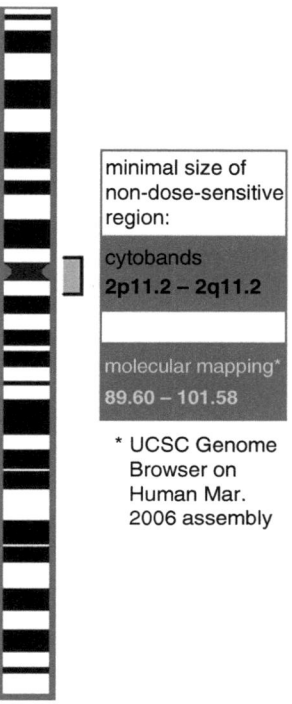

minimal size of non-dose-sensitive region:

cytobands
2p11.2 – 2q11.2

molecular mapping*
89.60 – 101.58

* UCSC Genome Browser on Human Mar. 2006 assembly

Table 6.2 Symptoms regularly observed in sSMC(2) cases with clinical symptoms (alphabetic order)

Symptoms	2p (%)	2q (%)
Dysmorphic face	50	83
Heart defect	–	16
Hypotonia	–	16
Kidney dysplasia/problems	25	16
Mental retardation	50	83
Microcephaly	–	50
Seizures	–	16

Cases distinguished by centromere-near partial trisomy of the short arm and the long arm according to three and six informative cases, respectively, collected in Liehr (2011a)

in partial trisomy distal from 2q11.2 and not in sSMC derived from the short arm of chromosome 2; that this could be specific is supported by case 02-W-IMB-q11.2/1-1 (Liehr 2011a), involving a patient with karyotype 46,XX,dup(2)(q11.2q14.2) and the same clinical sign.

6.2.2.1 Trisomy and Tetrasomy 2p or 2q

Partial trisomy or tetrasomy of the entire short arm or the entire long arm of chromosome 2 has never been reported.

6.2.3   Mosaicism

Mosaicism possibly having an influence on clinical signs and symptoms was reported in the sSMC(2) case 02-W-p10/1-1 (Liehr 2011a) . Here an sSMC(2) was first detected in a 6-year-old boy with psychotic illness, attention deficit, hyperactivity, stereotypic movements, mild mental retardation, and microcephaly. Later, the sSMC turned out to be maternally derived. The mother had as clinical signs only minor dysmorphism. The percentage of cells with sSMC was 54% in the mother and 44% in the son. However, only peripheral blood was examined, and it is well known that the degree of mosaicism may differ widely between different tissues. (Fickelscher et al. 2007).

6.2.4   Uniparental Disomy

No sSMC cases correlated with UPD 2 have been reported. Nonetheless, iUPD 2 might be a problem to be considered, as more than 30 UPD 2 cases have been reported in the literature (Liehr 2011c). Over 50% of these are connected with the activation of a recessive allele.

6.2.5   Case Report

Listed in Liehr (2011a) as 02-W-p11.2/1-2; reported by the mother of the now 14-year-old boy, with karyotype 47,XY,+min(2)(:p11.2->q11.1:)[11]/46,XY[10].

Our son was the product of a "very normal" pregnancy. He was born weighing 5 pounds 2 ounces through a caesarian section after I had been in labor for about 14 h, as I wanted a normal birth. A couple of years before, I had already given birth to my now 18-year-old daughter by caesarian section.

In the first months of his life we never saw anything abnormal about Carlos [Note: most of the children's names have been changed in accordance with their parents' wishes] except he would get sick a lot he had a very severe asthma. This went on for a couple of months. He was very underweight even though I was breast feeding. We started to notice about the age of 6–7 months that he would not sit or even stand alone. We started therapy which, according to my feeling has never really worked. Nonetheless, he started crawling at about 1 year and 2 months of age and he walked around the age of 1 year and 9 months. We figured the reason for this delay was because he was always sick and he used a lot of asthma medications and antibiotics.

Time went on and Carlos was still very short for his age, he still is. We tried growth hormones for this; of course it did not work. So at the age of 5 years, when he started kindergarten he could always fit in with other children because he was so small.

   The situation of Carlos has always been very hard to us as a family. It affects us as a couple and of course his sister. We got the diagnosis that Carlos had an sSMC derived from chromosome 2 when he was 9 years old; we just got the information from the MD that it was a rare chromosome disorder with unclear prognosis. At that time Carlos has already had a very severe speech delay, for that he is in speech therapy. He used to be on Ritalin, because teachers could not handle him being so hyperactive. But 2 years ago (at age of 12 years) we changed school and thank God his present teacher helped him to be off of it.

   Carlos has a very hard time with change. He cries a lot when he is taken somewhere he is not familiar with. A different bus, a different route to school or home he kind of has anxiety attacks. His routine has to be kept in order for him to be happy. He has improved some on his speech. I know he has learned a lot in school and actually school has helped him more than his therapies.

   He is a very lovely child. He participates in the Special Olympics every year. However, there he has a disadvantage because he competes with kids his own age and he is very small; he actually (at age of 14) looks like an 8 year old child.

   Carlos has a very good memory. He loves computers and can log onto webpage Nick or PBS (Play educational games) or Disney. He has matured in the last years but is still on diapers till this day. He does not wear one to school only when he gets home and for bowel movements only. We have tried everything in the book to resolve this, but have not been successful. He does not sleep by himself, is either with my husband and me or with his sister.

   Carlos has been a handful but we love our child and we want the best for him. God willing when he turns 15 he will start at a college to learn about computers even if he takes longer than normal, the school will have him there until needed. I hope his little size will not affect him going there. Maybe by then he would have stretched a little. Anyway, I only took him back once more.

6.3   Chromosome 3

6.3.1   Potentially Non-dose-sensitive Pericentric Region

More clinically normal than affected sSMC(3) carriers have been reported (Liehr 2011a). Possibly in accordance with this, the putative inert cytoband region is relatively large (3p12.2 to 3q13.1). Molecularly, only the 87.60–96.01 Mb region has been confirmed to be present in clinically normal sSMC carriers (Fig. 6.4).

   As previously mentioned for sSMC derived from chromosome 2, in sSMC(3) not only are partial trisomies tolerated, but partial tetrasomies are also tolerated, such as in cases 03-O-p12.1/1-1, 03-O-p12.1/2-1, and 03-O-p12.1/2-2 (Liehr 2011a).

Fig. 6.4 Right of the
ideogram of chromosome 3
drawn according to
Kosyakova et al. (2009) the
possibly non-dose-sensitive
pericentric region is
highlighted. The
corresponding cytobands
according to ISCN (*black
bracket*) and the molecular
mapping data (Mb) of the
region are given (*light-gray
square within black bracket*)

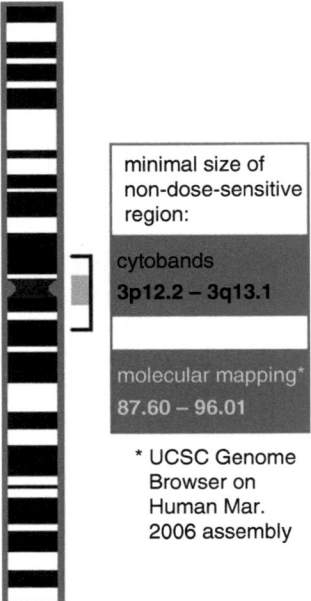

minimal size of
non-dose-sensitive
region:

cytobands
3p12.2 – 3q13.1

molecular mapping*
87.60 – 96.01

* UCSC Genome
Browser on
Human Mar.
2006 assembly

6.3.2   Clinical Signs

As only three euchromatic sSMC(3) cases with clinical signs have been reported
(cases 03-W-p12.1/1-1, 03-W-p12.1/2-1, and 03-CW-1; Liehr 2011a), it can only
be stated that mental retardation including developmental delay, and dysmorphic
features were present. No specific clinical signs due to imbalances of proximal
chromosome 3 materials are known.

6.3.2.1   Trisomy and Tetrasomy 3p or 3q

Partial trisomies or tetrasomies of the entire short arm or the entire long arm of
chromosome 3 are not compatible with life.

6.3.3   Mosaicism

Mosaicism, especially cryptic mosaicism (see Sect. 1.4.3), is present in most
sSMC(3) cases. A significant influence on clinical outcome is not yet obvious;
however, this may change when more cases are analyzed.

6.3.4   Uniparental Disomy

Overall, around ten UPD 3 cases are known. Among them there is only one sSMC case of a patient with the karyotype 47,XX,+min(3)(:p12.2->q10:) plus a maternal mixed hUPD/iUPD 3 (case 03-U-8; Liehr 2011a). The corresponding child was clinically affected, but it was not clear if this was due to the sSMC, a non-detected cell clone with karyotype 47,XX,+3, or partial iUPD 3.

6.4   Chromosome 4

6.4.1   Potentially Non-dose-sensitive Pericentric Region

As far as is currently known, the pericentric region of chromosome 4 has most probably no non-dose-sensitive euchromatic parts (Fig. 6.5). Two cases of sSMC(4) have even been reported with clinical symptoms and absence of any euchromatic imbalance induced by them: cases 04-W-p11/1-1 and 04-W-p11/1-2. However, in these cases it is most likely that the reported clinical problems were not due to the sSMC presence, as described in cases 04-U-2 and 04-U-7. The patient in case 04-U-7 had besides an sSMC(4) a homozygous mutation in the *methylmalonic aciduria and homocystinuria type C* (*MMACHC*) gene in 1q34.1, whereas for the patient in case 04-U-2 fragile X syndrome was diagnosed as primarily symptom causative (Liehr 2011a).

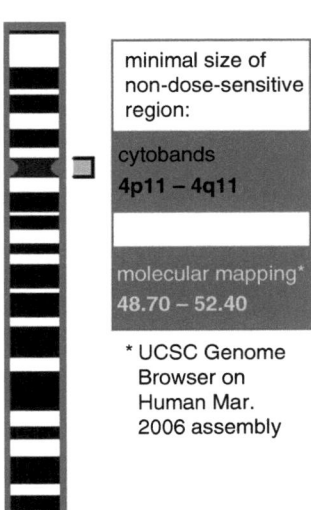

Fig. 6.5 Right of the ideogram of chromosome 4 drawn according to Kosyakova et al. (2009) the possibly non-dose-sensitive pericentric region is *highlighted*. The corresponding cytobands according to ISCN (*black bracket*) and the molecular mapping data (Mb) of the region are given (*light-gray square within black bracket*)

minimal size of non-dose-sensitive region:

cytobands
4p11 – 4q11

molecular mapping*
48.70 – 52.40

* UCSC Genome Browser on Human Mar. 2006 assembly

Table 6.3 Symptoms regularly observed in sSMC(4) cases with clinical symptoms (alphabetic order)

Symptoms	4 (%)
Clinodactyly	11
Dysmorphic face	89
Growth retardation	33
Mental retardation	77
Microcephaly	22

According to nine informative cases collected in Liehr (2011a)

The idea that there might nonetheless be a non-dose-sensitive pericentric region in chromosome 4 is underlined by case 04-W-p14/1-2 (Liehr 2011a). Here we see a partial trisomy of approx. 10 Mb in the proximal short arm of chromosome 4 and only relatively mild signs and symptoms. Yet, interpretation is made more difficult in this case as the sSMC(4) was present in only 60–70% of tissues studied.

6.4.2   Clinical Signs

The presence of an sSMC(4) has not been correlated with any specific clinical features as too few patients have been reported. However, dysmorphic features and mental retardation are seen most frequently (Table 6.3).

6.4.2.1   Trisomy and Tetrasomy 4p

According to Schinzel (2001), duplications of the whole short arm of chromosome 4 may arise primarily because of different types of familial rearrangements. As the observed pattern of dysmorphism, mental retardation, and heart defects are similar to those observed in cases with duplication exclusively of 4pter to 4p14, the critical region seems to be at the distal end of this chromosome arm (see Sect. 7.4). N.B.: Partial trisomy 4p can also be induced by an sSMC (case 04-W-p15.3/2-1; Liehr 2011a). Interestingly, no cases with tetrasomy 4p have been reported.

6.4.2.2   Trisomy and Tetrasomy 4q

Partial trisomies or tetrasomies of the entire long arm of chromosome 4 are not compatible with life.

6.4.3   Mosaicism

As mentioned already (see Sect. 6.4.1), mosaicism may play a role in some sSMC(4) cases, such as in case 04-W-p14/1-2 (Liehr 2011a). Owing to the limited number of reports, neither the influence of cytogenetically visible mosaicism nor the influence of cryptic mosaicism on clinical outcome can be estimated.

6.4.4   Uniparental Disomy

About ten cases of patients with maternal UPD 4 have been reported (Liehr 2011c). One of these patients is a carrier of an sSMC(4) and an additional chromosome 21; in this case, 04-U-1 (Liehr 2011a), the impact of partial trisomy 4, segmental UPD 4, and trisomy 21 cannot be distinguished. In about half of UPD 4 cases, activation of a recessive gene mutation took place (Liehr 2011c). Recently, one case of paternal UPD 4 was detected.

6.5   Chromosome 5

6.5.1   Potentially Non-dose-sensitive Pericentric Region

At least 18 Mb of the pericentric region of chromosome 5 seems to be potentially not dose-sensitive if it is present in three or even four copies; for the latter, see cases 05-O-p13.2/1-1, 05-O-p13.1/1-1, and 05-O-p10/1-1 (Liehr 2011a). Gain of cytobands 5p13.2 to 5q11.2, corresponding to 37.21–55.27 Mb, was without clinical consequences (Fig. 6.6) in at least 15 sSMC(5) cases.

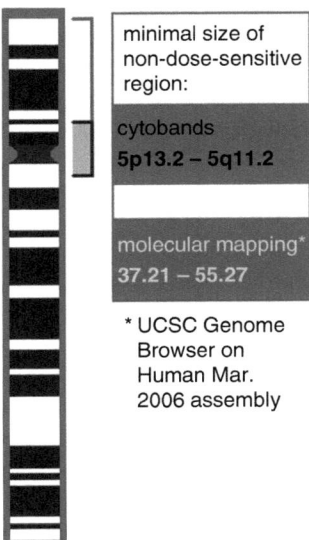

Fig. 6.6 Right of the ideogram of chromosome 5 drawn according to Kosyakova et al. (2009) the possibly non-dose-sensitive pericentric region is *highlighted*. The corresponding cytobands according to ISCN (*black bracket*) and the molecular mapping data (Mb) of the region are given (*light-gray square within black bracket*). The short arm is marked by a *gray half-bracket* as it may be exceptionally unproblematic if sSMC(5) is present in a mosaic

minimal size of non-dose-sensitive region:

cytobands
5p13.2 – 5q11.2

molecular mapping*
37.21 – 55.27

* UCSC Genome Browser on Human Mar. 2006 assembly

Table 6.4 Symptoms regularly observed in sSMC (5) cases with clinical symptoms (alphabetic order)

	5p (%)	5q (%)
Dysmorphic face	88	50
Growth retardation	44	50
Heart defect	44	50
Hypotonia	33	50
Macrocephaly	33	–
Mental retardation	55	100

Cases distinguished by centromere-near partial trisomy of the short arm and the long arm according to nine and two informative cases, respectively, collected in Liehr (2011a, b, c)

6.5.2 Clinical Signs

Typical signs of partial trisomy or tetrasomy close to the centromere of the short arm of chromosome 5 are dysmorphism and macrocephaly (Table 6.4). The latter has not been reported for imbalances close to the centromere of the long arm of chromosome 5. Besides, in sSMC(5) cases with clinical signs, heart defects, growth retardation, and hypotonia are regularly present.

6.5.2.1 Trisomy and Tetrasomy 5p

Trisomy of the short arm of chromosome 5 has been reported more than 20 times (Schinzel 2001; Liehr 2011a). Five cases are due to an sSMC(5) (Liehr 2011a), most others arose in connection with familial translocations (Schinzel 2001). The critical region for trisomy 5p syndrome seems to be located in the distal part of the short arm as the symptoms are similar to those seen in cases with pure trisomy 5pter to 5p13 (see Sect. 7.5). The most common features are mental retardation, facial dysmorphism, and hypotonia.

In the cases with tetrasomy 5p due to an sSMC, besides symptoms as in trisomy 5p, seizures were also present. Furthermore, life expectation may be significantly reduced (Schinzel 2001; Liehr 2011a).

6.5.2.2 Trisomy and Tetrasomy 5q

Partial trisomies or tetrasomies of the entire long arm of chromosome 5 have never been reported.

6.5.3 Mosaicism

There is only one case with trisomy of the short arm of chromosome 5 and no clinical symptoms have been reported (case 05-O-p15-33/1-1; Liehr 2011a). Mosaicism definitely plays a role, as the sSMC(5) was present in only 3% of the peripheral lymphocytes, hampering the establishment of trisomy 5p syndrome as reported before (see Sect. 6.5.2). Cases such as this have to be considered if trisomy 5p or tetrasomy 5p is detected prenatally in a mosaic state to provide correct genetic counseling (Fig. 6.6).

6.5.4 Uniparental Disomy

Three cases with UPD 5 have been reported, all of them leading to the activation of a recessive gene mutation (Liehr 2011c). No sSMC(5) case with UPD is known.

6.5.5 Chromosomes 1/5/19

See Sect. 6.1.6.

6.6 Chromosome 6

6.6.1 Potentially Non-dose-sensitive Pericentric Region

According to cytogenetic data, the potentially non-dose-sensitive pericentric region of chromosome 6 spans at least cytobands 6p12.1 to 6q11.1. Molecular confirmation remains to be done and only the centromeric region from 58.40 to 63.40 Mb is for sure a non-dose-sensitive region (Fig. 6.7).

6.6.2 Clinical Signs

Only six cases of patients with sSMC(6) with clinical signs have been reported, four involving only short arm material and two involving exclusively long arm material (Liehr 2011a). Even though from Table 6.5 there might be differences between these two groups of patients, presently it can only be stated for sure that dysmorphism and mental retardation are present in sSMC(6) cases. Possible microcephaly is indicative more for partial trisomy of the short arm of chromosome 6 (Table 6.5).

6.6.2.1 Trisomy and Tetrasomy 6p

There are no reports of sSMC including the whole short arm of chromosome 6. However, trisomy 6pter to 6q12 has been reported, leading to microcephaly, mental

Fig. 6.7 Right of the ideogram of chromosome 6 drawn according to Kosyakova et al. (2009) the possibly non-dose-sensitive pericentric region is *highlighted.* The corresponding cytobands according to ISCN (*black bracket*) and the molecular mapping data (Mb) of the region are given (*light-gray square within black bracket*)

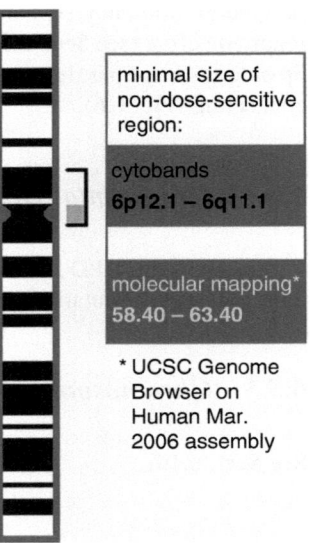

minimal size of non-dose-sensitive region:

cytobands
6p12.1 – 6q11.1

molecular mapping*
58.40 – 63.40

* UCSC Genome Browser on Human Mar. 2006 assembly

Table 6.5 Symptoms regularly observed in sSMC(6) cases with clinical symptoms (alphabetic order)

Symptoms	6p (%)	6q (%)
Dysmorphic face	75	100
Growth retardation	25	–
Heart defect	–	50
Hypotonia	–	50
Macrocephaly/overgrowth	–	50
Mental retardation	25	100
Microcephaly	75	–

Cases distinguished by centromere-near partial trisomy of the short arm and the long arm according to four and two informative cases, respectively, collected in Liehr (2011a)

retardation, seizures, and heart defect (Schinzel 2001). Tetrasomy 6p seems to be a lethal condition.

6.6.2.2 Trisomy and Tetrasomy 6q

Complete partial trisomy or tetrasomy 6q is not viable.

6.6.3 Mosaicism

Three of nine sSMC(6) patients with clinical abnormalities have cryptic mosaicism, leading in part to partial tetrasomy or even hexasomy. Because of limited case numbers, the impact of these gains of copy number is hard to address.

6.6.4   Uniparental Disomy

UPD of chromosome 6 should be considered carefully in clinical cases with sSMC derived from chromosome 6, as paternal UPD 6 is known as one of the imprinting disorders, leading to transient neonatal diabetes (OMIM #601410). Approximately 80% of the reported UPD 6 cases are of paternal origin (Liehr 2011c). One case of maternal UPD 6 and one case of paternal UPD 6 has been observed with sSMC(6) (Liehr 2011c). Whereas transient neonatal diabetes is the main feature in paternal UPD 6, maternal UPD 6 can be correlated with intrauterine growth retardation. Additionally, iUPD 6 leading to activation of a recessive point mutation has been reported (Liehr 2011c).

6.6.5   Case Report

Provided by Unique; reported by the mother of a 22-year-old man with karyotype mos 47,XY,+min(6)(:p21.1->q21:)/46,XY.

Craig [Note: most of the children's names have been changed in accordance with their parents' wishes] has blessed our lives for 22 years now. At his birth, June 1987, he was healthy and seemed normal although he was small for a full term baby (5 lbs 2 oz or 2.3 kg). Nothing alarmed us until about 2 years of age when he was not walking or talking. At that time, his doctor suggested a blood test which revealed an abnormal karyotype. In 12 out of 13 cells he had extra chromosome material that was assumed to be a piece of the upper arm of chromosome 18. At that time, a blood test on his father and me showed normal chromosome patterns (Fig. 6.8).

As a baby, Craig was quiet but had an easy smile and loved to be held. He finally began walking around the age of 2, and began talking around the age of 4 years with the help of a speech pathologist. After his diagnosis he was placed in Early Intervention, a way for the school to aid his education, and remained in special education classes until he turned 19.

In the year 2000, after significant progress in genetic research, we did another study and found that Craig's first diagnosis had been incorrect. The new study revealed a very different chromosome result: 47,XY,+der(6)del(6)(p21.1)del(6) (q21).ish der(6)(wcp6+)[15]/46,XY[14]. It was mosaic – only 6 out of 13 cells tested

Fig. 6.8 Craig at the age of
22 years (copyright Unique)

showed the anomaly – hence, I like to think that in another 10 years, it might go away altogether![1] *It was also noted in the report that the anomaly occurred post conception.*

As Craig grew, he showed signs of anxiety and sometimes would hide under his school desk. Today Craig struggles with anger typically because he does not understand how to deal with emotions. Throughout his life, Craig has received psychological treatment mostly for the use of medicines to help him control emotions. He was diagnosed with Asperger autism and does respond to autism spectrum behavior treatments.

Craig has a small stature and a very young face. He loves to read and is adept at playing video games. He is somewhat uncoordinated at riding bicycles and longs to ride a skate board like a professional, but is content to pretend using one with Wii. He has lived on his own in an apartment but requires help. He has a job which he loves as a part-time aide at the airport. I asked him about his favorite memories and he says that he remembers most fondly many summers at a Mother-Child camp we attended on the Oregon Coast. There he was accepted without question by everyone. As I asked him these questions, he proudly showed me that he had completed one face of a Rubik's Cube! Craig is loved by his family and is especially close to his grandfather who loves to make Craig laugh.

6.7   Chromosome 7

6.7.1   Potentially Non-dose-sensitive Pericentric Region

The potentially non-dose-sensitive pericentric region of chromosome 7 spans at least cytobands 7p11.2 to 7q11.22, corresponding to molecular positions 56.45–67.00 Mb (Fig. 6.9). However, this suggestion is based on only a single case (07-O-p11.2/1-1; Liehr 2011a). Additionally, two more sSMC(7) cases without clinical symptoms have been reported; however, they were not characterized in detail for the exact sSMC sizes. As the size estimation of the putative non-dose-sensitive pericentric region of chromosome 7 is based on only one case, which additionally is mosaic (the sSMC is present in less than 50% of peripheral blood cells; see Sect. 6.7.4), one has to deal extremely carefully with these data. Additionally, a possible UPD 7 has always to be considered (see Sect. 6.7.4).

---

[1]*Comment by the author*: Note that only peripheral blood cells were studied here, which derive from bone marrow, a fast dividing tissue. If other cells of Craig's body were studied, one would most likely find the sSMC in almost every cell, as these cells do not proliferate so much and are not prone to loss of the sSMC like blood cells. Even if no sSMC(6) is present in peripheral blood, Craig's clinical problems would, because of that, persist, like in PKS cases (see Sect. 5.3).

Fig. 6.9 Right of the ideogram of chromosome 7 drawn according to Kosyakova et al. (2009) the possibly non-dose-sensitive pericentric region is *highlighted*. The corresponding cytobands according to ISCN (*black bracket*) and the molecular mapping data (Mb) of the region are given (*light-gray square within black bracket*)

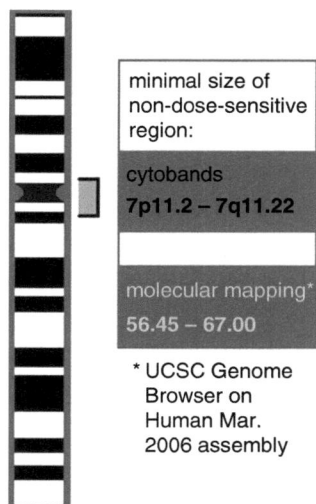

minimal size of non-dose-sensitive region:

cytobands
7p11.2 – 7q11.22

molecular mapping*
56.45 – 67.00

* UCSC Genome Browser on Human Mar. 2006 assembly

6.7.2 Clinical Signs

Dysmorphism, macrocephaly, and mental retardation are present in sSMC(7) with clinical signs, irrespective of whether the imbalance is derived from the long arm or the short arm (Table 6.6). Hypotonia might be more related to partial trisomy close to the centromere of the long arm. Overall, signs and symptoms are non-specific, also due to the number of low cases.

Table 6.6 Symptoms regularly observed in sSMC(7) cases with clinical symptoms (alphabetical order)

Symptoms	7p (%)	7q (%)
Autism	–	16
Dysmorphic face	67	100
Heart defect	–	16
Hypotonia	–	33
Macrocephaly	67	50
Mental retardation	100	83

Cases distinguished by centromere-near partial trisomy of the short arm and the long arm according to three and six informative cases, respectively, collected in Liehr (2011a)

6.7.2.1 Trisomy and Tetrasomy 7p

As in sSMC(6), no sSMC derived from the short arm of chromosome 7 are known. However, there are familial duplications of the entire short arm of chromosome 7, leading to severe clinical features, including a markedly reduced lifespan. Partial tetrasomy 7p seems to be lethal, as this has not yet been reported (Schinzel 2001).

6.7.2.2   Trisomy and Tetrasomy 7p

No case of complete trisomy 7q has been reported. The same is true for tetrasomy 7q.

6.7.3   Mosaicism

Mosaicism (case 07-O-p11.2/1-1; Liehr 2011a) as well as cryptic mosaicism (case 07-W-p11.2/1-2; Liehr 2011a) could have an influence on the clinical outcome of sSMC(7) carriers. More case reports are needed to delineate the meaning of mosaic karyotypes.

6.7.4   Uniparental Disomy

Six sSMC(7) cases have been reported to be correlated with maternal UPD 7. Overall, more than 100 UPD 7 cases are known from the literature (Liehr 2011c), most of which are maternal UPD. The latter are correlated with SRS (OMIM #180860), an "imprinting disorder." No paternal UPD 7 syndrome is known.

According to OMIM (#180860) SRS is clinically heterogeneous. Characteristic features are severe intrauterine growth retardation, poor postnatal growth, cranio-facial features such as a triangular-shaped face and a broad forehead, body asymmetry, and a variety of other minor malformations. The phenotypic expression changes during childhood and adolescence.

About 15 of the reported UPD 7 cases are correlated with the activation of recessive mutations due to (segmental) iUPD 7; in eight of these cases the gene causative for cystic fibrosis (cystic fibrosis transmembrane conductance regulator, CFTR gene in 7q31.2) is affected.

6.7.5   Case Report

Listed in Liehr (2011a) as case 07-W-p11.1/1-1; reported by the mother of the now 10-year-old girl.

Rebecca [Note: most of the children's names have been changed in accordance with their parents' wishes] was born in Brussels, after normal pregnancy and birth. Directly after birth the midwife noticed a wry neck, which "appears regularly" in newborn and would "outgrow" with time. The latter was a word I had to be prepared to hear more often from now on. Breast feeding was difficult, but I was used to that, as with Rebecca's elder sister we had the same problem. However, after not drinking anything for 2 days finally the pediatrician was wondering and prescribed a sonography of the throat. Together with this investigation a sonography of Rebecca's head was done, as well. So, 3 days after her birth I was told that my daughter had a brain defect. I will remember this conversation for ever. I was told that Rebecca lacks the corpus callosum, a structure of the brain which connects and coordinates both cerebral hemispheres. The neurologist told us that many children lack corpus callosum and 30% of them develop normally. About the other 70% …. I did not ask for them. He told that Rebecca would possibly never learn to walk or talk. In between Rebecca acquired a laryngomalacia, which caused problems during breathing, but which (certainly!) would disappear with age. Having these diagnoses Rebecca and I were released from the hospital 10 days after birth.

What came now was like hell for us. I had to go to work 8 weeks after birth and my husband took over care for our daughter – that was how we intended to do. However, Rebecca could not suck and had problems to breathe. It was not until a surgery of her larynx at the age of 6 months that she started – according to my opinion – to live, as she recovered her breath and could focus on other things. Now finally she showed her first smiling, which we have been waiting for such a long time. Still she did not want to drink and she was alimented by a nose probe. Then, with a lot of patience and supported by a Castillo Morales therapist my husband taught her to eat puree.

At the age of 1 year Rebecca moved to Germany with us, my husband went back to work and I took over child care. We started with early intervention at 17 months of age, as she had substantial developmental delay. It was clear that she did not belong to the above mentioned 30%. Still I needed several years until I was able to see that clearly. I needed time until I could use the world "disabled" for my daughter and still I have not accepted the handicapped state of her completely.

At the beginning Early Intervention was difficult, as Rebecca had difficulties to accept new reference persons. Step by step she adopted the terrific offerings and tried them out. She crawled over towered mattresses, learned to sit, to pull up and to stand. She played in sand, used the sliding board, put all available things into her mouth, swashed in water, built towers, played manhunt, arranged her first puzzle, set the table with plastic dishes and fed her doll. Rebecca made substantial progress in her development, which were amazing to us.

The social pediatric center caring for us suggested a cytogenetic investigation, when our daughter was 2 years. A human geneticist told us thereafter the result: in half of her cells Rebecca had a marker chromosome. The MD suggested at first a possible cultural artifact, however, a few months later an additional blood test confirmed the result. Further studies revealed that mine and the chromosomes of my husband were normal. When our daughter was around 3 years we were referred to Dr. Liehr, Jena, by our human geneticist, as a specialist for this rare chromosomal condition. Further blood tests should clarify the origin of the sSMC and give possibly hints on the reason for Rebecca's developmental delay. As taking blood is difficult for Rebecca (and all people involved in that procedure, especially the mother), I want to spare that to her. Thus, blood is always only taken under general anesthetic, which is necessary from time to time to install a new grommet.

The first results from Jena came and suggested a uniparental disomy (UPD) 7 could be the reason for Rebecca's problems. If so our daughter would have inherited one chromosome #7 from one parent twice. What a devastating message. The necessary additional studies including parental blood analysis did luckily not confirm this idea. What was clear: the sSMC was derived from chromosome #7; accordingly the marker chromosome did not lead to gain of genetic relevant material. Thus, the meaning of that chromosomal aberration remained unclear.

At 4 years of age Rebecca was presented to a specialist for rare syndromal diseases. The date was set long time in advance, finally it took place. I had big hopes and expectations. I conceived this specialist would examine Rebecca exactly and then tell me: she has the xyzsyndrome. I hoped he would tell me exactly in which phase of embryonic development this syndrome evolved. Then I could go to the library of the university, certainly would find a whole shelf with professional literature for the xyz syndrome, and could finally learn how Rebecca's many different abnormalities go together and how her further development will be. But it came completely different. Rebecca was examined tenderly and intensively. The specialist really found additional abnormalities not yet recognized by anyone (alterations in pigmentation in the skin on her tummy, creases on the foot). But these and all other abnormalities would be "mild" and would not point towards a known syndrome. What was for sure: Rebecca did not inherit these abnormalities form her parents. This was a release for me as this finally took away my feelings of guilt and also my sorrows about my older daughter. For her there is no enhanced risk to have a handicapped child.

Nonetheless, I felt infinitely empty after this discussion; again no diagnosis, no clarity. But after some time I was able to see the positive aspects: Rebecca's future is open and not written in any book. I am not to blame for my daughter's the obstruction. And very slowly I calmed down.

Meanwhile Rebecca went to a special Kindergarten with a small therapeutic pedagogic group. Troublesome she had been learning to walk and still moved loose and without great endurance. Her speech development was markedly delayed, which was at least in parts due to her moderate deafness diagnosed at 2 years of age. She got hearing aid devices for both ears and she got special training courses at home and in kindergarten. As Rebecca did not talk a word we were recommended to offer her supported communication. I was against it for long time as I thought she would never learn to talk. However, the contrary was the case. I started to address her with simple sign language and Rebecca picked up on that immediately. The morning when she was able to decide herself by means of sign language what she wanted to have on her buttered bread (i.e., cheese or sausage) I had tears in my eyes. We obtained a "talker", tinkered picture book and step by step Rebecca anticipated the new possibilities of communication. At Christmas 2005 Rebecca said "baba" and clearly intended to say "Papa", the German word for "dad"! At Whitsun 2007 Rebecca said "Mama" (German for "mum") – what a great day!

Besides, there would be many more things to report – about double kidney, tympanoplasty, seizures, etc. but I do not want to go into more details here. Still we are supervised by our human genetic center nearby with periodical consultations for follow-up. However, at present there is nothing else to do to find out more about Rebecca's problems. So here we have a little rest now.

With 6 years Rebecca entered a special school for hearing-impaired persons. Supported by sign language Rebecca finally found the speech. Today she speaks full sentences, understandable at least for those persons dealing every day with her. Still her language improves; she knows the entire ABC and can "gesticulate" all letters. Due to problems with fine motor skills it is hard for her to write. She knows the figures until 10, is very interested in her environment, tries to understand procedures and relationships and can memorize in detail things like, e.g., different species of frogs.

The daily life with Rebecca can be very tiring. When awake (during night she luckily sleeps about 10 h without break!), she is 100% present and there is not a second of break. She has a "bullhead" and a strong will, recognizes no endangerments, and has problems to accept rules. She is in favor of having contacts with adults and has no friends of her age. It is only her sister she plays with nowadays already for longer.

Recently we made a bike ride with the whole family. On a meadow two fringes were curried and prepared for the hack. Rebecca did watch this scene very long and it was hard to make her move away. At home she told: "The horses were riding with the children in the forest" – what a sentence! Experiences like that provide strength to cope with every day's life and to face new challenges courageously.

6.8   Chromosome 8

6.8.1   Potentially Non-dose-sensitive Pericentric Region

sSMC(8) are the most frequently observed non-acrocentric-derived marker chromosomes. About 15 sSMC(8) cases have been reported without clinical symptoms. According to them, the potentially non-dose-sensitive pericentric region of chromosome 8 spans subbands p11.21 to q11.21. Only region 42.50–48.30 Mb has been molecularly confirmed, by case 08-O-p11.21/3-1 (Liehr 2011a) (Fig. 6.10). In cases of clinically normal patients – apart from case 08-O-p23.1/1-1 (Liehr 2011a), which is discussed later (see Sect. 6.8.3) – only partial trisomies have been reported.

6.8.2   Clinical Signs

sSMC derived from the short arm of chromosome 8 with clinical signs differ from those derived from the long arm of chromosome 8 (Table 6.7). Autism,

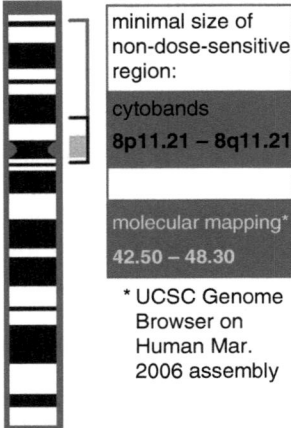

Fig. 6.10 Right of the ideogram of chromosome 8 drawn according to Kosyakova et al. (2009) the possibly non-dose-sensitive pericentric region is *highlighted*. The corresponding cytobands according to ISCN (*black bracket*) and the molecular mapping data (Mb) of the region are given (*light-gray square within black bracket*). The short arm is marked by a *gray half-bracket* as it may be exceptionally unproblematic if sSMC(8) is present in a mosaic

Table 6.7 Symptoms regularly observed in sSMC(8) cases with clinical symptoms (alphabetic order)

Symptoms	8p (%)	8q (%)
Autism	25	–
Dysmorphic face	50	86
Hypermobility of joints	25	–
Hypertelorism	17	28
Macrocephaly	25	–
Mental retardation	83	57
Microcephaly	–	14
Overgrowth	17	14

Cases distinguished by centromere-near partial trisomy of the short arm and the long arm according to 12 and seven informative cases, respectively, collected in Liehr (2011a)

hypermobility of the joints, and macrocephaly are typically found in partial trisomy 8p cases; microcephaly was present only in partial trisomy 8q cases. Dysmorphism and hypertelorism were more frequent in proximal partial trisomy 8q, whereas mental retardation was slightly more frequent in proximal partial trisomy 8p.

6.8.2.1   Trisomy and Tetrasomy 8p

Trisomy of the entire short arm of chromosome 8 may occur in connection with familial rearrangements or because of an sSMC (Schinzel 2001; Liehr 2011a). Typical clinical features are short stature, a characteristic dysmorphic face, hypertelorism, scoliosis, mental retardation, agenesis of corpus callosum, cardiac malformations, macrocephaly or hydrocephalus, and other malformations.

Partial tetrasomy of the short arm of chromosome 8 has been reported only in connection with isochromosome 8p formation. The clinical signs are similar to those of partial trisomy 8p and include hydrocephalus, agenesis of corpus callosum, cardiac malformations, typical facial dysmorphism, and mental retardation (Schinzel 2001; Liehr 2011a).

6.8.2.2 Trisomy and Tetrasomy 8q

Recently one mosaic case with trisomy 8q was reported (Wood et al. 2008), which led to growth retardation, dysmorphic features, and termination of the pregnancy. No tetrasomy 8q is known.

6.8.3 Mosaicism

Mosaicism seems to play a role in sSMC(8) cases. All clinically normal carriers had mosaicism of cells with and without sSMC in the tissue studied. Many clinically abnormal sSMC(8) carriers were mosaic and/or cryptic mosaic.

Interestingly, there is one case with almost complete mosaic trisomy and tetrasomy of the short arm of chromosome 8, which is not connected with clinical signs: case 08-O-p23.1/1-1 (Liehr 2011a). The carrier had the sSMC(8) in approximately one third of her peripheral blood cells. Thus, the short-arm region of chromosome 8 is marked with a gray bracket in Fig. 6.10.

6.8.4 Uniparental Disomy

Fewer than ten UPD 8 cases have been reported. No sSMC case is among them. However, three UPD 8 cases were detected because of activation of a recessive gene mutation (Liehr 2011c).

6.8.5 Case Report

Listed in Liehr (2011a) as 08-W-p11.21~11.22/1-1; reported by the father of the now 18-year-old boy with karyotype 47,XY,+r(8)(::p11.21~11.22->q11.1::).

Our son was born 2 weeks prematurely in 1993 – APGAR score 9 having icterus. He refused to nurse (except from the bottle) and suffered from colic the first year. He was delayed to late-normal for most motor development milestones: walking on his own at 20 months. His eye contact and social connection regressed after his second birthday. His speech was also late-normal – he had some language but stagnated between the age of 2 and 5 with little new language acquisition. Although we felt it was obvious that he was delayed, his playgroup teacher (twice a week for from age 3 to 5 years) told us she was sure he was fine and that he "just needed more time". At 5 he received the diagnosis "pervasive developmental disorder – not otherwise specified = PDD-NOS", a kind of atypical autism. Around that time cytogenetics was done and an sSMC(8) was found; karyotype: 47,XY,+r(8)(::p11.21~11.22->q11.1::).

We started a special behavioral therapy (applied behavior analysis = ABA) which made a great difference. He only then started to learn the basics for school.

He went to the local school with an aide through the sixth grade and now attends a special education school for learning disabilities and goes unassisted. He travels independently to and from school. His IQ test results have varied widely from low 70s in the "Wechsler intelligence scale for children = WISC" to 97 (Kaufman assessment battery for children = K-ABC). Physical appearance is normal. He still has reduced abstract thinking ability, thus, difficulty with math, history, geography, time, money, reading etc. His experiential memory is good. His favorite activities are TV (music channels), videos, iPod, movies, and eating.

Our son has had digestive problems since birth and recurring constipation/ diarrhea. Since the age of 5 he has been on a strict casein (milk)/gluten free diet which has greatly reduced his digestive issues and improved his sleep. Before the diet he awoke every night between 2 and 5 AM and could not get back to sleep – after the diet he awoke in the night on average only once a week.

He is average size for his age with the only unusual feature – larger than normal big toes on both feet. Also he has had chronic lip herpes and is less coordinated than average, but still can play school sports. He is socially shy, does not really understand most social cues and is slower to process language. This results in a fear that he will act inappropriately and he avoids most social situations unless feeling very comfortable as in the family.

We are confident that he will be able to find appropriate work and live independently with some supports in place.

6.9   Chromosome 9

6.9.1   *Potentially Non-dose-sensitive Pericentric Region*

Owing to the large heterochromatin block in the centromeric region, the potentially non-dose-sensitive pericentric region of chromosome 9 spans at least the molecular region 42.96–70.50 Mb and cytobands 9p13.1 to 9q21.12 (Fig. 6.11). Cryptic mosaicism (cases 09-O-p12/3-1 to 09-O-p12/3-2 and 09-O-p11.1/2-1; Liehr 2011a) and noncryptic mosaicism (case 09-O-p12/6-1; Liehr 2011a) leading to partial tetrasomies in clinically normal persons can also be viable.

6.9.2   *Clinical Signs*

Only a few sSMC(9) cases with clinical signs, for which sufficient clinical information was provided, are available in the literature. No typical features for these

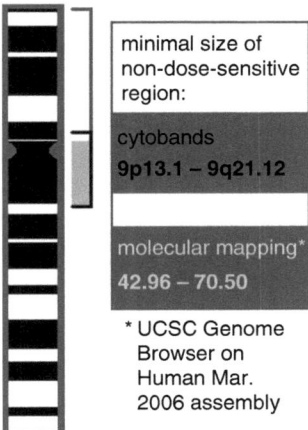

minimal size of
non-dose-sensitive
region:

cytobands
9p13.1 – 9q21.12

molecular mapping*
42.96 – 70.50

* UCSC Genome
Browser on
Human Mar.
2006 assembly

Fig. 6.11 Right of the ideogram of chromosome 9 drawn according to Kosyakova et al. (2009) the possibly non-dose-sensitive pericentric region is *highlighted*. The corresponding cytobands according to ISCN (*black bracket*) and the molecular mapping data (Mb) of the region are given (*light-gray square within black bracket*). The short arm is marked by a *gray half-bracket* as it may be exceptionally unproblematic if sSMC(9) is present in a mosaic

patients can be identified at this point. There was only mental retardation as a common clinical sign in cases 09-W-IMB-q12/1-1, 09-CW-1, and 09-CW-3 (Liehr 2011a).

6.9.2.1 Trisomy and Tetrasomy 9p

Partial trisomy of the short arm of chromosome 9 is known as a recognizable syndrome, with more than reported 100 cases (Schinzel 2001). Intrauterine and postnatal growth retardation, microcephaly, mental retardation, dysmorphic features, and malformations of the fingers and toes are typically observable. An isochromosome 9p can accompany partial trisomy and tetrasomy 9p. However, in most such cases, isochromosome 9p leads to partial tetrasomy 9p, leading to similar, but often more obvious symptoms than those in partial trisomy 9p (Schinzel 2001). Overall, more than 40 cases with isochromosome 9p have been reported (Liehr 2011a).

6.9.2.2 Trisomy and Tetrasomy 9q

No corresponding cases are known, even though rarely mosaic trisomy 9 was reported (Schinzel 2001).

6.9.3   Mosaicism

As shown in Fig. 6.11, mosaicism is an important feature in cases with sSMC(9). There are now three case reports for carriers of isochromosomes 9p without clinical signs (cases 09-O-pter/1-1 to 09-O-pter/1-3; Liehr 2011a). Most probably they have in most of their body tissues the otherwise disease-causing sSMC(9) only in a subset of the corresponding cells. Owing to the limited number of reported cases, there are no further clues to the influence of mosaicism in sSMC(9) cases, even though most clinically normal patients are mosaics.

6.9.4   Uniparental Disomy

More than 12 cases with maternal UPD, two cases with paternal UPD, and two cases with UPD of unclear origin have been reported. Chromosome 9 does not seem to be subject to imprinting; however, iUPD and activation of recessive genes was reported for at least two cases. For one sSMC(9) case a maternal UPD 9 was reported; however, the clinical significance is not clear in that case (Liehr 2011a).

6.9.5   Case Report

Provided by Unique; reported by the mother of a girl with karyotype mos 47,XX,+i (9p)/46,XX.

When you first hear the cries of your newborn baby, you have this overwhelming feeling that the whole purpose of your existence has been waiting for that exact moment your new baby was born. It is impossible to even describe the immense feeling of love, responsibility and commitment you have for something that is so new, and so small.

It is the same when you hear those first words telling you your child has a chromosomal disorder. That moment will stay with you forever. Only, it is not a moment of sheer happiness, it is pure disbelief and sorrow. No words can ever describe the sadness you feel for your baby when you are told they may not be able to enjoy life's experiences we all so desire to achieve.

That is how it was for me and my whole family when my fifth child Alison [Note: most of the children's names have been changed in accordance with their parents' wishes] was diagnosed with mosaic tetrasomy 9p due to marker chromosome presence. There was no information on this syndrome and the doctor could not tell us what to expect. They did not know what life span she would have, what degree of intellectual disability to expect, or what her quality of life would be when she was older. Had the information been available at that time, I doubt I could have absorbed it anyway. Little did we know then that this syndrome would by no means stop Alison from achieving her own purpose, her own self satisfaction and her own individual status in life!

In saying that, Alison has had to work hard. Learning was often tedious, monotonous and exhausting. She struggled through her milestones. When other children her age were walking, she was crawling. When they were talking, she was managing one word sentences. And when they were laughing and enjoying conversations with friends, Alison was still learning to put sentences together. Eventually though, Alison, too was doing all those things. She may have been older, but she got there in the end (Fig. 6.12).

The most difficult time for Alison was in her early teenage years. For the first time in her life she really felt alone and singled out. She became depressed and spent most of her time hiding in her room. I decided it was time to once again search for another child with the same syndrome. With the help of a wonderful website called Unique, the rare chromosome disorder support group and their magazine, we finally found another child with mosaic tetrasomy 9p. Alison had been alone with her syndrome for 15 years and she finally had her first contact with a little boy with the same syndrome. Although he lived in a different part of the world and because he was only a baby, communication could only be via email with his parents Alison felt an incredible belonging to that child. That little boy saved her life, just by being who he was. Alison did not feel alone anymore and was able to go back to being the fighter we all knew her to be.

Through Alison's eyes, I have learnt so many things. She is so much stronger than I am in so many ways. She has taught me to never give up, just find another way. She believes life is about what you make it for yourself, not what others think it should be. She has showed all of us that being what society calls "normal" is not what makes you happy. It's who you are inside that really matters.

Fig. 6.12 Alison at the age of 17 years (copyright Unique)

6.10　Chromosome 10

6.10.1　*Potentially Non-dose-sensitive Pericentric Region*

According to eight ssMC(10) cases and one case with a similar chromosomal imbalance due to a duplication near the centromere, cytobands 10p21.1 to 10q11.22 are potentially non-dose-sensitive (Liehr 2011a). Molecularly, the inert region was defined to span at least positions 34.40–44.45 Mb (Fig. 6.13).

All nine cases without clinical symptoms mentioned before led at most to proximal partial trisomy of chromosome 10. No cases with cryptic mosaicism leading to partial tetrasomies have been reported for ssMC(10).

Fig. 6.13 Right of the ideogram of chromosome 10 drawn according to Kosyakova et al. (2009) the possibly non-dose-sensitive pericentric region is *highlighted*. The corresponding cytobands according to ISCN (*black bracket*) and the molecular mapping data (Mb) of the region are given (*light-gray square within black bracket*)

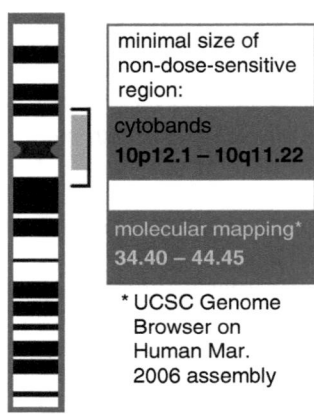

minimal size of non-dose-sensitive region:

cytobands
10p12.1 – 10q11.22

molecular mapping*
34.40 – 44.45

* UCSC Genome Browser on Human Mar. 2006 assembly

6.10.2　*Clinical Signs*

In Table 6.8 the clinical signs seen in five cases with larger ssMC(10) are listed. The symptoms reported for the four cases with pericentromeric partial trisomy 10p are relatively uniform, but also nonspecific. For one case with pericentromeric partial trisomy 10q, overgrowth is reported; however, in eight cases with similar imbalances not caused by ssMC(10) this feature was not observed (cases 10-W-IMB-q11.1/1-1 to 10-W-IMB-q11.1/7-1 and 10-W-IMB-q11.2/1-1; Liehr 2011a).

6.10.2.1　Trisomy and Tetrasomy 10p

According to Schinzel (2001), more than 50 cases of partial trisomy 10p have been reported, mainly induced by familial chromosomal rearrangements. This imbalance causes prenatal and postnatal growth retardation, microcephaly, facial dysmorphism, malformations of the extremities, mental retardation, and a wide spectrum of

Table 6.8 Symptoms regularly observed in sSMC(10) cases with clinical symptoms (alphabetic order)

Symptoms	10p (%)	10q (%)
Dysmorphic face	100	–
Growth retardation	100	–
Heart defect	25	–
Hypotonia	50	–
Mental retardation	100	–
Overgrowth	–	100

Cases distinguished by centromere-near partial trisomy of the short arm and the long arm according to four informative cases and one informative case, respectively, collected in Liehr (2011a)

additional possible clinical features. Mosaic tetrasomy 10p was once prenatally reported as being associated with marked sonographic signs (Wu et al. 2003b).

6.10.2.2   Trisomy and Tetrasomy 10q

Trisomy or tetrasomy of the long arm of chromosome 10 seems not to be viable.

6.10.3   Mosaicism

Interestingly, apart from "normal mosaics" such as 47,XN,+mar/46,XN, no cryptic mosaics have been reported for sSMC derived from chromosome 10. This finding may change, with more cases becoming available or pointing toward lethality of proximal partial tetrasomies of chromosome 10.

6.10.4   Uniparental Disomy

One sSMC case was connected with maternal UPD 10 (case 10-U-2, Liehr 2011a). As the corresponding pregnancy was electively terminated, the clinical impact remains unclear. Overall, four maternal and one paternal UPD 10 cases are known (Liehr 2011a).

6.10.5   Case Report

Provided by Unique; reported by the parents of a boy with karyotype 46,XY,dup(10) (q11.2q22.3)de novo. N.B.: Even though this case has no sSMC(10) but a centromere-near duplication, the clinical outcome would have been similar in a case with karyotype 47,XY,+r(10)(::p11.1->q22.3::).

Our son, Dear Son (DS), has a rare genetic condition known as "partial proximal trisomy 10q". This means that part of the long arm of chromosome 10 is duplicated, so there are three copies of the duplicated section rather than two. There is no syndrome name for what our son has and there are fewer than 10 reported cases in the world. Not much is known about the disorder, so we are learning as we go.

DS was born at full term with slight complications (he could not expel the mucus from his lungs) and was rushed to neonatal intensive care. Several heart defects were then discovered that had not been detected by prenatal ultrasound. The most severe was a coarctation of the aorta. In addition, he had a hypoplastic left ventricle and atrial and ventricular septal defects. At 3 days old, our son had heart bypass surgery to correct his coarctation. After this, we expected him to develop normally. His small head was concerning but this might be hereditary, so we did not worry much about it.

However, as DS grew, he struggled to do things that our older child had done with ease. He did not meet certain milestones and was significantly delayed in others. He seemed "floppy". He struggled to chew food, often pooling it in his mouth, and vomited. He struggled with tasks that required fine motor capabilities, did not talk on schedule, developed a "lazy eye" and had below average control of his trunk muscles. His weight was low and at various points he was termed "failure to thrive". He has always been slight of frame.

We began to see several therapists, both through the county's Early Intervention program and privately. It was suggested that DS had had a "hard start" due to his heart surgery and things would level out. That was not the case. We struggled on for about a year and a half until a pediatrician recommended genetic testing and in short order we found out about DS's condition. We had declined the opportunity to participate in a genetic study after DS's heart surgery, since it meant yearly visits out of state. How we wish we had!

However, everything now started to make sense. Most of DS's difficulties could be explained and, in time, we have accepted his condition as part of who he is. We no longer tried to "fix" our son, but concentrate on helping him to achieve his fullest potential.

We continued, though, to struggle with feeding issues, ongoing since birth. After trying a behavior-based therapy and having countless medical people tell us "When he's hungry he will eat", or "He eats and drinks – what's the problem?", or "He's just small", we enrolled DS in a 2-month-long day patient feeding clinic at a hospital for special needs children. We feel this is one of the best things we have done for DS medically.

As a result of evaluations, we learned that DS had extensive oral motor issues, including not knowing how to chew and process food properly. In addition, DS has a high-arched palate and dental abnormalities (large spaces between teeth, not enough teeth and oddly shaped teeth). He also has texture and temperature aversion to some foods.

Through a program of intensive oral motor exercises (Beckman) and the introduction of high-calorie puréed foods, liquid supplementation, positive behavior reinforcement, and eating on a schedule according to a specific protocol, DS not only learned to help feed himself but also gained 6-plus pounds in 2 months. DS can eat non-puréed food as a side dish or snack, but since he struggles to consume enough calories to sustain himself, he will eat the bulk of his meals puréed for now.

The program was a great success. Not only did DS gain weight, but also additional skills. DS can now pick up a loaded spoon and fork and use them to feed himself, whereas before he was feeding himself almost entirely by hand. He can drink from a regular straw and regulate his eating and drinking much better. We try to keep his diet free from most processed foods as he has reactions at times to food that include lip swelling and rashes on his face. Test results have been inconclusive or yielded conflicting results. Tests to determine whether or not he has an immunoglobulin (IgG) deficiency were normal. During this past year, we also discovered that DS's tonsils are swollen to about 2 on a scale of 4 – with 4 indicating a need for removal.

Apart from medical issues, our son is an active and happy child who likes to play on the computer and read books. He can navigate some children's websites using a two-button mouse, can use an iPad, and enjoys simple applications on a touch screen phone. He enjoys going to the swimming pool (though cannot swim) and swinging. He likes to stack blocks and knock them down and is learning to tell time. He enjoys school and is excited to ride the bus each morning.

DS spends most of his school day in a self-contained classroom at our base school. He receives occupational and speech therapies at school, and also takes speech therapy privately. He can read on a first grade level and knows many words on sight. He tries to sound out words he does not know, indicating phonemic awareness. He enjoys reading books or looking at the pictures. He identifies strongly with certain cartoon or entertainment characters and has become interested in more mature ones as he has aged, though his interests at age 11 are those of a much younger child. He has a great memory. At the age of 10, DS lost his first tooth and started to teeth big time.

He is learning basic addition skills, using manipulatives and numbers. He is working on expressing his comprehension via the "w questions". DS does not like to hold a pencil and cannot write. He resists using a stamp with his name on and labels. There is a disconnect between whatever he is holding in his hand (unless it involves a computer or book) and his brain and this makes anything but scribbling with a crayon very difficult. DS is also learning to recognize money and coins, and do simple sorting tasks. Though smart in some ways, his level of functioning is not very high.

When DS is motivated, he tries hard. He enjoys helping around the house and receiving praise for a job well done. Both at home and at school, he is learning "life skills" –like recycling, helping to unload the dishwasher, and collecting and sorting books. He requires adult assistance with most routine tasks. He cannot yet dress or undress himself independently and is not toilet trained. He can wash his hands with the right kind of faucet but will not do so independently. He must be prompted to do even the most basic of self care tasks, but can follow directions.

Our son's receptive language skills are much better than his expressive skills, though his communication has improved greatly since attending the feeding clinic, with its intensive regimen of oral motor exercises and our follow up. He also has a wonderful sense of humor, evident by his sly smile, an occasional joke, and mischievous behavior at times!

DS has some compliance issues and his behavior centers around falling to the ground and resisting physically when he does not want to do something. Prior he ran away but no longer does that. He understands consequences and a reward for good behavior is a good way to motivate him. Recently, his behavior has greatly improved with use of a behavior chart and firmly stated expectations.

DS has many sensory issues. He does not have a good sense of himself in space and frequently moves from side to side or twirls around in large open spaces or on unfamiliar ground. He seeks out certain textures and fabrics for comfort. He has favorite stuffed animals, several quite large, which he likes to put on top of himself when he falls asleep. He likes "fidget" toys and has one in his hand almost constantly. He does not have a good sense of safety, and we have to watch him so that he does not walk out into the street, for example.

Though the first years of DS's life were tough on us, we try to take the challenge of having a child like DS in our stride. We have another child, who is 2 years older, and so try to take the whole family's needs into consideration. Still, there are probably many things that we do not do simply because it is too hard. DS cannot deal with the noise or commotion of certain events so we do not attend many fireworks displays, for example. And DS does not have the same stamina as other kids. But we try to be as normal a family as possible and to give our kids family experiences that are fun and rewarding.

Because of DS we have learned the value of real friendship and kindness and what is really important in life. DS is a joy and a dear child and we would not trade him for anything in the world.

6.11   Chromosome 11

6.11.1   Potentially Non-dose-sensitive Pericentric Region

Two nonmosaic cases with an sSMC(11) (cases 11-O-p11.2/1-1 and 11-O-p11.1/2-1) and two carriers of intrachromosomal duplications (cases 11-O-IMB-p11.2/1-1 and 11-O-IMB-p11.2/2-1) provide evidence that the potentially non-dose-sensitive pericentric region of this chromosome spans at least cytobands 11p11.2 to 11q12.2, i.e., 45.60–60.23 Mb (Fig. 6.14). Even mosaic partial tetrasomy of part of this region is tolerated in case 11-O-p11.1/2-1 (Liehr 2011a).

6.11.2   Clinical Signs

Imbalances of the pericentric region of chromosome 11 cause no really specific clinical features. Mental retardation and dysmorphism are the most frequent

Fig. 6.14 Right of the ideogram of chromosome 11 drawn according to Kosyakova et al. (2009) the possibly non-dose-sensitive pericentric region is *highlighted*. The corresponding cytobands according to ISCN (*black bracket*) and the molecular mapping data (Mb) of the region are given (*light-gray square within black bracket*)

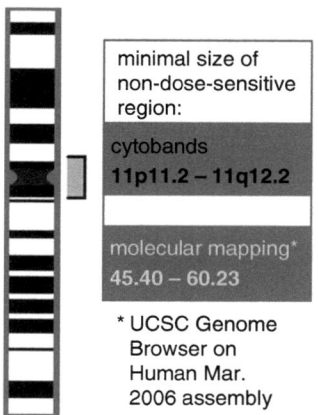

minimal size of non-dose-sensitive region:

cytobands
11p11.2 – 11q12.2

molecular mapping*
45.40 – 60.23

* UCSC Genome Browser on Human Mar. 2006 assembly

abnormalities, followed by growth retardation, heart defects, and hypotonia (Table 6.9).

Table 6.9 Symptoms regularly observed in sSMC(11) cases with clinical symptoms (alphabetic order)

Symptoms	11 (%)
Dysmorphic face	60
Growth retardation	40
Heart defect	40
Hypotonia	40
Macrocephaly	20
Mental retardation	80
Seizures	20

According to five informative cases collected in Liehr (2011a)

6.11.2.1 Trisomy and Tetrasomy 11p

As especially the short arm of chromosome 11 is subject to imprinting (see Sect. 6.11.4), the clinical outcome of a gain of copy number of the short arm of chromosome 11 depends partly on the parental origin of the imbalance. The reported patients all had familial chromosomal rearrangements. For maternal origin only one case has been reported, with unclear impact of the gain of copy number of the short arm of chromosome 11 on the clinical outcome (Schinzel 2001). For paternal origin, the affected persons showed features such as those observed in BWS (OMIM #130650) – see Sect. 6.11.4). Partial trisomy of the whole short arm of chromosome 11 has not been observed in connection with and caused by an sSMC(11). Also, partial tetrasomy 11p seems to be not compatible with life.

6.11.2.2 Trisomy and Tetrasomy 11q

No whole-arm trisomies or tetrasomies of chromosome 11 have been reported.

6.11.3   Mosaicism

Mosaicism and cryptic mosaicism have been reported in sSMC(11) cases with and without clinical abnormalities. An obvious influence on the clinical outcome has not been reported.

6.11.4   Uniparental Disomy

UPD 11 has not been seen together with an sSMC derived from chromosome 11. Interestingly, in contrast to other chromosome-specific UPD, in chromosome 11 paternal UPD (more than 100 cases) is more frequently present than maternal UPD (four cases).

   Chromosome 11 is subject to imprinting; thus, paternal UPD 11 leads to BWS (OMIM #130650) and maternal UPD 11 leads to SRS (OMIM #180860; see Sect. 6.7.4). BWS, most often caused by segmental paternal UPD 11p, leads to the following clinical features: prenatal and postnatal overgrowth, macrocephaly, large tongue, dysmorphism, normal or mildly subnormal mental development, and prenatally a large placenta (OMIM #180860). Additionally, at least two cases have been reported with activation of a recessive allele leading to other rare diseases due to iUPD 11 (Liehr 2011c).

6.12   Chromosome 12

6.12.1   Potentially Non-dose-sensitive Pericentric Region

The potentially non-dose-sensitive pericentric region of chromosome 12 spans at least cytobands 12p12.2 to 12q12, corresponding to molecular positions 28.47–39.90 Mb (Fig. 6.15).

   Mosaicism can have an influence on the clinical outcome (see Sect. 6.12.3); nonetheless in clinically normal patients no proximal partial tetrasomy and no cryptic mosaicism have been found (Liehr 2011a).

6.12.2   Clinical Signs

For partial gain of copy number in chromosome 12 close to the centromere, no really unique and/or specific features have been reported (Table 6.10). Regularly found are mental retardation and dysmorphism; growth retardation, microcephaly, and heart defects may also be present. Strikingly, hypotonia is only seen in cases with partial trisomy near the centromere of the short arm rather than the long arm of chromosome 12.

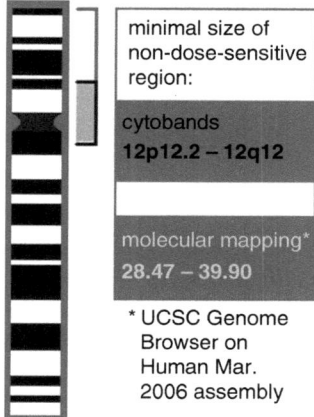

minimal size of
non-dose-sensitive
region:

cytobands
12p12.2 – 12q12

molecular mapping*
28.47 – 39.90

* UCSC Genome
Browser on
Human Mar.
2006 assembly

Fig. 6.15 Right of the ideogram of chromosome 12 drawn according to Kosyakova et al. (2009) the possibly non-dose-sensitive pericentric region is *highlighted*. The corresponding cytobands according to ISCN (*black bracket*) and the molecular mapping data (Mb) of the region are given (*light-gray square within black bracket*). The short arm is marked by a *gray half-bracket* as it may be exceptionally unproblematic if sSMC(12) is present in a mosaic

Table 6.10 Symptoms regularly observed in sSMC(12) cases with clinical symptoms (alphabetic order)

Symptoms	12p (%)	12q (%)
Dysmorphic face	33	83
Growth retardation	33	50
Heart defect	33	33
Hypotonia	67	–
Mental retardation	67	83
Microcephaly	33	33
Seizures	–	16

Cases distinguished by centromere-near partial trisomy of the short arm and the long arm according to three and six informative cases, respectively, collected in Liehr (2011a, b, c)

6.12.2.1 Trisomy and Tetrasomy 12p

Trisomy of the short arm of chromosome 12 has been reported in approx. 30 cases. Typically these patients have a facial appearance resembling that of Down syndrome; besides, one can see in this group male genital hypoplasia, short and broad hands and fingers, clubfoot, and mental retardation (Schinzel 2001).

Tetrasomy 12p is consistent with PKS (see Sect. 5.3). As shown in Fig. 6.15, as in cases with isochromosomes 8p (see Sect. 6.8.2) and 9p (see Sect. 6.9.2) so in isochromosome 12p a mild(er) phenotypic expression can be found due to mosaicism (case 12-Wpks-1, Liehr 2011a).

6.12.2.2 Trisomy and Tetrasomy 12q

These conditions seem to be nonviable.

6.12.3 Mosaicism

Mosaicism is present in sSMC(12) with and without clinical symptoms. Cryptic mosaicism has been seen in only two cases with clinical signs (cases 12-W-p11.1/2-1 and 12-CW-2; Liehr 2011a). However, as mentioned in Sect. 6.12.2.1, mosaicism can influence the clinical outcome especially in cases with isochromosome 12p.

6.12.4 Uniparental Disomy

Six UPD 12 cases are known, four with maternal UPD, and one each with paternal and unclear parental origin of UPD. In one case maternal UPD 12 was present together with an sSMC(12), but no clinical symptoms were present. Neither imprinting nor activation of recessive genes has been seen in any of these five cases. But the latter is in principle possible with iUPD 12.

6.12.5 Case Report

See Sect. 5.3.3.

6.13 Chromosome 13

6.13.1 Potentially Non-dose-Sensitive Pericentric Region

Chromosome 13 is an acrocentric chromosome; for this kind of autosome it is commonly agreed that the short arms do not contain any relevant genetic information, apart from the NOR. Thus, the putative non-dose-sensitive pericentric region of chromosome 13 includes at least the whole short arm. Additionally, part of the long arm close to the centromere seems to be inert even to partial tetrasomy (Fig. 6.16).

As shown in Fig. 6.16, this region corresponds overall to cytobands 13pter to 13q12.2; molecular mapping data are not available. However, the possibly non-dose-sensitive region in the long arm of chromosome 13 is based on only one case

Fig. 6.16 Right of the ideogram of chromosome 13 drawn according to Kosyakova et al. (2009) the possibly non-dose-sensitive pericentric region is *highlighted*. The corresponding cytobands according to ISCN (*black bracket*) and the molecular mapping data (Mb) of the region are given (*light-gray square within black bracket*)

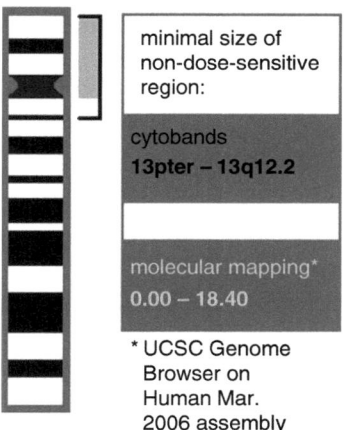

minimal size of non-dose-sensitive region:

cytobands
13pter – 13q12.2

molecular mapping*
0.00 – 18.40

* UCSC Genome Browser on Human Mar. 2006 assembly

(13-O-q12.2/1-1; Liehr 2011a), and the data should be interpreted accordingly with care.

6.13.2 Clinical Signs

There are no clinically informative aberrant sSMC(13) cases. However, there are seven cases of proximal long arm partial trisomy 13q12~13 due to intrachromosomal duplications which had in common short stature, microcephaly, dysplastic ears, strabismus, epicanthic folds, down-slanting palpebral fissures, short mandible, and mild to severe mental retardation (see cases 13-W-IMB-q12/1-1 to 13-W-IMB-q12/1-5, 13-W-IMB-q13/1-1, and 13-W-IMB-q13/1-2; Liehr 2011a).

6.13.2.1 Trisomy and Tetrasomy 13q

Complete trisomy 13 leads to the well-known Patau syndrome, which can also be caused by the presence of an isochromosome 13q in a numerically normal karyotype (Schinzel 2001). Complete tetrasomy 13q has never been reported.

6.13.3 Mosaicism

Mosaicism and cryptic mosaicism can be present in sSMC(13) cases. No correlation can be made at this time, because of small case numbers.

6.13.4   Uniparental Disomy

Five maternal UPD 13, seven paternal UPD 13, and two maternal or paternal UPD 13 cases have been reported (Liehr 2011c). Three of four cases with iUPD 13 had homozygous mutations in the connexin 26/GJB2 gene; however, there was no case with sSMC(13).

6.13.5   Chromosomes 13/21

Most human chromosome pairs have, for unknown reasons, a specific centromeric DNA sequence which can be used to distinguish each of them from the others by means of molecular cytogenetics. However, chromosome 13 is the most important exception to this rule, as is shares its centromeric sequence with chromosome 21, and has no other individual centromeric DNA sequence (see Sect. 1.3, Fig. 1. 2). In over 100 sSMC(13) and sSMC(21) cases it was impossible to specify the chromosomal origin other than as these two chromosomes (Liehr 2011a). Especially, as most of these sSMC obviously contain no euchromatic material, it is impossible to define them unambiguously. Most of these sSMC do not lead to any clinical problems. For those which are found in clinically abnormal persons, it is highly probable that the sSMC itself is not disease-causative, but just an incidental finding.

6.14   Chromosome 14

6.14.1   Potentially Non-dose-sensitive Pericentric Region

As mentioned for chromosome 13, chromosome 14 is also an acrocentric chromosome. Thus, the entire short arm is non-dose-sensitive. Overall, cytobands 14pter to 14q11.2, i.e., 0.00–19.88 Mb, are potentially non-dose-sensitive. Two mosaic cases (14-O-q11.2/2-1 and 14-O-q11.2/3-1) and two nonmosaic cases (14-O-q11.2/4-1 and 14-O-q11.2/5-1) have been reported, supporting this suggestion (Fig. 6.17) (Liehr 2011a).

At the most partial trisomy 14q near the centromere has, however, been seen in clinically normal cases, but no cryptic mosaicism or tetrasomy has been seen.

6.14.2   Clinical Signs

Among 12 sSMC(14) cases with clinical signs, dysmorphic features and mental retardation are most often reported. As listed in Table 6.11, the remaining clinical

Fig. 6.17 Right of the ideogram of chromosome 14 drawn according to Kosyakova et al. (2009) the possibly non-dose-sensitive pericentric region is *highlighted*. The corresponding cytobands according to ISCN (*black bracket*), and the molecular mapping data (Mb) of the region are given (*light-gray square within black bracket*)

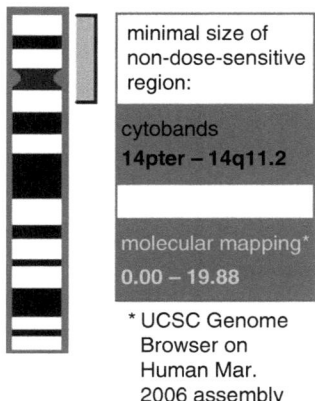

minimal size of non-dose-sensitive region:

cytobands
14pter – 14q11.2

molecular mapping*
0.00 – 19.88

* UCSC Genome Browser on Human Mar. 2006 assembly

Table 6.11 Symptoms regularly observed in sSMC(14) cases with clinical symptoms (alphabetic order)

Symptoms	14q (%)
Clinodactyly	16
Dysmorphic face	75
Hearing deficit	25
Heart defect	25
Hypotonia	25
Mental retardation	58

According to 12 informative cases collected in Liehr (2011a)

features are heterogeneous and hardly specific, even though hearing problems, which could be a specific symptom, are found in a quarter of the cases.

6.14.2.1 Trisomy and Tetrasomy 14q

Trisomy of chromosome 14 is only viable if it is present in a mosaic with a normal cell line. According to Schinzel (2001), approx. 20 such patients have been reported. Pure trisomy 14 is lethal and tetrasomy is not known.

6.14.3 *Mosaicism*

In clinically normal patients, cytogenetically visible mosaicism but no cryptic mosaicism has been described. Also in sSMC(14) with clinical signs, cryptic mosaicism is rather rare. Clinically, the meaning of mosaicism in sSMC(14) is still unresolved.

6.14.4 Uniparental Disomy

Chromosome 14 is subject to maternal and paternal imprinting. Temple syndrome (see OMIM *605636 and #176270) is due to maternal UPD 14; paternal UPD 14 also leads to typical clinical features, but does not have its own denomination apart from "paternal uniparental disomy 14 syndrome" (OMIM #608149). Most of the more than 60 reported cases of UPD 14 are characterized by inheritance of a Robertsonian translocation from one parent, or have a de novo inverted duplication chromosome 14q.

Paternal UPD 14 syndrome presents with skeletal abnormalities, joint contractures, dysmorphic facial features, and (develop)mental retardation and is often lethal in the intrauterine period (OMIM #608149). Typically in paternal UPD 14, placentomegaly is observed (Ogata et al. 2008). Among the more than 35 reports of paternal UPD 14, there is one case with an sSMC(14) (Liehr 2011c).

"Maternal uniparental disomy 14 syndrome" (Temple syndrome) (see OMIM *605636 and #176270) is variable in expressivity. Patients with Temple syndrome can be confused with patients with TS (see Sect. 5.5) or PWS (see Sect. 6.15.4); the phenotypic spectrum includes nearly normal phenotypes. Clinical features are hypotonia (89%), small hands (89%), low birth weight (85%), short stature (85%), early onset of puberty (85%), facial dysmorphism (68%), obesity (43%), and developmental delay (37%) (Ogata et al. 2008).

6.14.5 Case Reports

Report A: see Sect. 4.2. Personal experience 4.

Report B: Provided by Unique; reported by the mother of a 7-year-old female with karyotype 47,XX,+mar(14)(pter->q12)[24]/46,XX[26].

It seems hardly possible that my daughter has just turned 7. Minutes ago I went to see Hannah [Note: most of the children's names have been changed in accordance with their parents' wishes] as she lay in the bed where she spends most of her day. She did not say anything, because she cannot. But she looked at me so intently. She knows who we are, what goes on around her, and even understands what is said to her. I know this because when you ask her to hold her head up, she tries to, and when you ask her to find Mama, she will turn her eyes and look, if she feels like it. Only if she feels like it, though – that is how I know that she really is a 7-year-old (Fig. 6.18).

We had a bit of a scare before Hannah was born, because she had a two-strand umbilical cord rather than the normal three-strand, but the doctors kept a close eye on her, and everything seemed normal when she was born. Her challenges only became apparent over the next few months as she failed to gain weight and failed to hit the normal developmental benchmarks. Ultimately, a genetic test revealed a bit of extra material on the 14th pair of chromosomes.

Fig. 6.18 Hannah at the age
of 7 years (copyright Unique)

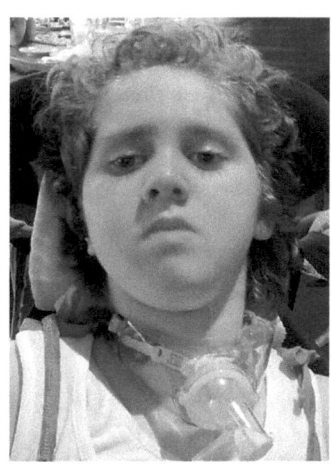

We are grateful for the extra bit that did not materialize: the part that would have given her a major heart defect and the other part that would have profoundly warped her face. The extra that is there is enough to silence her and keep her from walking, rolling over on her own, or even closing her epiglottis to swallow, because it has taken from her the ability to coordinate her movements.

She has other diagnoses that likely stem from the trisomy issue, but may be unrelated. These include developmental delay, scoliosis, cerebral palsy, failure to protect her airway, chronic reactive airways, chronic lung disease secondary to chronic aspiration, chronic urinary tract infections, failure to thrive, laryngomalacia, physical and neuromuscular delays, and windswept hip syndrome. We deal with over 15 different doctors' offices, at least half a dozen medical supply and equipment suppliers, and three different types of therapists.

Hannah is not coordinated, but she is strong – not a good combination when the random movements of her tongue sometimes place it between her teeth when her jaw muscles randomly clamp down. Regular botulinum injections to her jaw muscles prevent her from reducing her tongue to hamburger. The worst times are when the botulinum is starting to wear off, but before she can have another injection. Many times during that period we have to use sedatives to protect her from hurting herself.

Because she is non-verbal and has to be suctioned almost constantly, Hannah needs to be watched 24 hours a day. She has no regular sleep pattern, so we never know when she will be awake or asleep. We have had 24-hour nursing for Hannah for just over the last 2 years, following a 58-day stay in the hospital when she received her tracheostomy. She has only been hospitalized once since then. We have had one other emergency visit since then, but they knew we had nursing care so they sent us home with instructions for the nurses and did not keep her in the hospital. Prior to that, she has been hospitalized 22 times. Having a nurse with her all the time has really made a big difference in her health.

It would be easy to just view Hannah in terms of her diagnoses and her health problems, and those things do occupy a lot of our thoughts. But she is so much more than that. Even though she does not talk to anyone (her vocal cords are normal so it seems to be a muscle control issue), she still communicates with us. With her eyes and actions she tells us that she hurts, and makes us wish she could tell us where. She tells us that she wants to tell us something, and you can see her intense struggle to get it across somehow. Sometimes you can tell that if her muscles would cooperate, she would be laughing, and she manages to almost smile. And when you blow kisses at her the movements of her lips really look like she is trying hard to blow them back. Those times are the best.

It is hard to get it across in words alone, but anyone who has spent some time with Hannah has no doubt that she is a fully aware, physically challenged, deep-souled little girl. Hannah has her favorite people, including one of our fill-in nurses. We do not blame her – he is so kind and gentle with her, almost like a grandfather would be. Not long ago he filled in a shift after we had not seen him for a month or two. When he got here Hannah took one look at him and then turned her head and refused to look at him or even acknowledge him again for about an hour. At that point it seems she was satisfied that he had learned his lesson and she began to pay attention to him. The same thing happened again a month or two ago when he filled in again.

Another one of her favorite nurses can calm her easily by crawling up in bed with her, and the two of them watch TV together, Hannah cradled in the nurse's arms as if they were on a bobsled. Many times Hannah will even fall asleep there because she is so calm and relaxed.

Here are two statements of reality, both of which are true: We ache for what might have been, wondering what she would be like if she had not been put in this situation (it's especially hard when we see other "normal" girls her age). We love her just as she is and would not change a thing.

Here are two more equally true statements: we are exhausted and drained, and we are so honored to have the responsibility for this special little person.

Because Hannah's condition is so rare, there is no medical database that will enable us to predict anything about her prognosis. We have found just a few people who have a similar mix of mosaic and partial trisomy 14. Our "benchmark" T-14 buddy is now 22 years old and doing very well. But we still know that Hannah may live a month, a year, a decade, or 80 more years. She may improve, or she may not. We hope she will but we do not know. But parents of "normal" kids do not really know either. They just have the illusion of certainty. We have simply realized that certainty is always an illusion, and thus we take each moment and treasure it.

6.14.6   Chromosomes 14/22

For chromosome 14 the same holds true as was reported for chromosome 13 in Sect. 6.13.5. Chromosome 14 shares its commercially available centromeric sequence with chromosome 22. In contrast to chromosomes 13 and 21, chromosomes 14 and 22 can be distinguished from each other, as chromosome 22 has another, individual centromeric sequence; however, the latter is not commercially available. More than 60 sSMC(14) and sSMC(22) cases have been reported (Liehr 2011a). As in chromosomes 13 and 21, these sSMC(14) and sSMC(22) do not contain euchromatin and normally do not lead to clinical problems.

6.15 Chromosome 15

6.15.1 Potentially Non-dose-sensitive Pericentric Region

Chromosome 15 is an acrocentric chromosome, so its entire short arm is non-dose-sensitive (see Sect. 6.13.1). Additionally, cytobands 15pter to 15q11.2, i.e., 0.00–21.05 Mb, are potentially non-dose-sensitive (Fig. 6.19). For this region, partial trisomies (cases 15-O-q11.2/4-1 to 15-O-q11.2/5-1), tetrasomies (cases 15-O-q11.2/1-1 to 15-O-q11.2/1-10), and even hexasomies (cases 15-O-q11.2/3-1 to 15-O-q11.2/3-3; Liehr 2011a) have been reported, which can all be tolerated, even if they are present in all cells studied.

Larger imbalances would only be inert if the sSMC(15) was in a mosaic with a normal cell line (gray half-bracket in Fig. 6.19) such as in cases 15-O-q11.2~12/1-1 to 15-O-q11.2~12/1-5, 15-O-q12/1-1 to 15-O-q12/1-2, 15-O-q13/1-1 to 15-O-q13/3-1, and 15-O-q13.1/1-1 (Liehr 2011a; see Sect. 6.15.3).

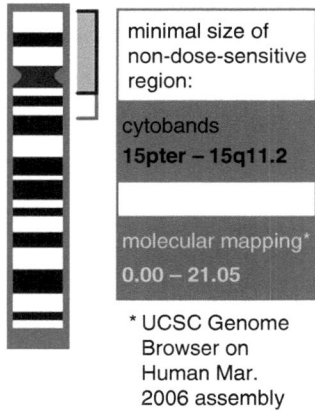

Fig. 6.19 Right of the ideogram of chromosome 15 drawn according to Kosyakova et al. (2009) the possibly non-dose-sensitive pericentric region is *highlighted*. The corresponding cytobands according to ISCN (*black bracket*) and the molecular mapping data (Mb) of the region are given (*light-gray square within black bracket*). The small part marked by a *gray half-bracket* highlights a region which may be unproblematic if sSMC(15) is present in a mosaic

Recent research has partly elucidated the formation of ssMC(15). LCR regions along the long arm of chromosome 15 cause at least five breakpoints or breakpoint cluster regions, leading to ssMC(15) of different sizes (Parokonny et al. 2007; Wang et al. 2008; Kleefstra et al. 2010).

6.15.2   Clinical Signs

According to Battaglia (2008), patients with larger ssMC(15) display distinctive clinical findings, and a tetrasomy 15q syndrome can possibly be defined – see also OMIM #608636. As summarized in Table 6.12, this concept can be confirmed and even refined when analyzing the data summarized in Liehr (2011a). Overall, ssMC(15) spanning 15pter to 15q12 have less severe phenotypes than larger ones including 15pter to 15q14, even though the spectrum of aberrations can overlap. Most important, autism and seizures appear (almost) exclusively in larger ssMC(15).

If the ssMC(15) with clinical signs induces only proximal partial trisomy, then the clinical signs seem to be similar, and not even less expressed (Liehr 2011a).

6.15.2.1   Trisomy and Tetrasomy 15q

Trisomy 15, including mosaic trisomy 15, seems to be lethal prenatally or perinatally (Schinzel 2001). There is only one viable exception, due to a translocation 46,X,der (X)t(X;15) and subsequent X-inactivation of the derivative X chromosome (Stankiewicz et al. 2006). Tetrasomy 15 has never been reported.

Table 6.12 Symptoms regularly observed in ssMC(15) cases with clinical symptoms (alphabetic order)

Symptoms	15pter-15q12 (%)	15pter-15q14 (%)
Autism	–	43
Dysmorphic face	18	6
Growth retardation	3	3
Heart defect	1	1
Hypotonia	7	5
Hymelanose Ito	–	3
Macrocephaly	1	–
Mental retardation	95	100
Microcephaly	–	2
Seizures	7	36

Cases distinguished by size of the ssMC according to 224 informative cases (74 spanning 15pter-15q12 and 150 for 15pter-15q14) collected in Liehr (2011a)

6.15.3 Mosaicism

Mosaics and cryptic mosaics are present in sSMC(15) with and without clinical problems. There seems to be a tendency toward more severe clinical features when the sSMC(15) leads to (mosaic) proximal partial hexasomy (cases 15-W-q13/2-1 to 15-W-q13/2-4 and 15-W-q13/6-1; Liehr 2011a) compared with those with partial tetrasomy. Additionally, imbalances normally leading to clinical aberrations can be tolerated in rare cases if a normal cell line is present – gray half-bracket in Fig. 6.19 (see Sect. 6.15.1).

6.15.4 Uniparental Disomy

Chromosome 15 is present in one maternal and one paternal copy in every cell of all healthy humans. In the case of maternal UPD 15 PWS (OMIM #176270) and in the case of paternal UPD 15 AS (OMIM #105830) is the consequence. Maternal UPD 15 has been reported approx. 450 times and paternal UPD 15 has been reported over 100 times. Among these cases, an sSMC(15) was present seven times in AS and 32 times in PWS (Liehr 2011c). Interestingly, in one maternal UPD 15 case, an sSMC derived from the X chromosome was detected (case 15-P-2; Liehr 2011a). PWS and AS are not only due to UPD; point mutations, imprinting defects, and microdeletions in the PWS/AS critical region of chromosome 15 are some of the other possible disease-causing mechanisms. Accordingly, in those sSMC cases with AS and PWS, a UPD 15 was only proven in three and 17 cases, respectively. Besides, one AS and two PWS cases with (heterochromatic) sSMC had a microdeletion in the PWS/AS critical region of chromosome 15. Thus, exclusion of a PWS/AS critical region microdeletion in the case of a prenatally detected de novo sSMC(15) seems to be indicated (Liehr et al. 2005; Liehr 2011a).

The clinical features of PWS, also called Prader–Labhart–Willi syndrome, are diminished fetal movements, muscular hypotonia as a newborn, short stature, hypogonadotropic hypogonadism, small hands and feet, scoliosis, mental retardation (mild 63%; moderate 31%; severe 6%), and later in life obesity due to hyperphagia, beginning between 6 months and 6 years of age. Also present in 75% of patients is low pigmentation (OMIM #176270; Kunze 2009).

Angelman syndrome (AS), formerly also called happy puppet syndrome, is characterized by a typical facial appearance (microbrachycephalia), mental retardation, severe limitations in speech and language, ataxic movement or balance disorder, hypotonia (93% of patients during their whole life), characteristic abnormal behavior, including a friendly and happy mood, and seizures (80%), the latter beginning between 6 and 42 months. Often patients are blond and blue-eyed (OMIM #105830; Kunze 2009).

6.15.5   Case Reports

Report A: See Sect. 4.2. Personal experience 1.

Report B: Provided by Unique; reported by the mother of a 17-year-old boy with karyotype 48,XY,+inv dup(15)(pter->q12::q12->pter)x2 de novo.

I sailed through a trouble-free pregnancy with no sickness, the only issue being that at my 20-week scan, a "black hole" showed up on the baby's head. The general practitioner reassured me it was probably nothing.

At 36 weeks the baby was still breech, so he was delivered at 38 weeks by elective Caesarean section. Colin [Note: most of the children's names have been changed in accordance with their parents' wishes] was born with a hole at the base of his spine so he was taken straight to X-ray. We were told it was a sacral dimple and nothing to worry about. Then at 3 days old, doctors discovered he had a heart murmur, which was very worrying: the mere mention of the word "heart" made us panic. We were taken to the specialist heart hospital in London to have the chambers of his heart investigated and were told he had a small ventricular septal defect that would close by the time he was 6 months old – and it did.

Colin slept a lot as a baby and most of the time I had to wake him up to feed him. He was, however, very colicky for about 4 months. At his 8-month clinic check he failed all the tests, and then failed them again a month later. He was then referred to our local hospital for blood tests. He was assessed at our local child development centre because of his global developmental delay and we finally got the diagnosis 2 weeks after his first birthday. Colin was diagnosed with inverted duplication of chromosome 15q12. He has two identical marker chromosomes, giving a 48-chromosome karyotype. We were devastated.

We were given cups of tea and details about Unique, a small support group for people affected by a rare chromosome disorder. The pediatrician gave us a copy of the lab report and sent us home to read it through and highlight anything we did not understand. The assessment that day was rescheduled for the following week.

Our first questions were "How long will he live, 5 years, 10 years, 20 years?" "Will he walk?" "Will he talk?" The pediatrician told us honestly that she did not know, only that sometimes children with a lowered immunity can get infections and pneumonias which can lead to death. We respected her answers.

At the time of diagnosis Colin could only lie on the floor. He was not able to roll over, sit, stand or crawl. He was also very floppy. We were referred to all sorts of different professionals: physiotherapy, occupational therapy, speech therapy and portage service (home-based early intervention). Suddenly our lives became an endless round of appointments. It was like being on a roller coaster ride that kept going round and round, stopping occasionally, only to start again. By the time Colin was 18 months old, the help from the physiotherapist and occupational therapists was starting to improve his life. He started to crawl. He spent a long time trying to pull himself up the furniture and eventually succeeded and started cruising, until after a year, at 3, he started to walk unaided. If he fell he could not understand how to stand up again and it took him 3 or 4 months to work this out.

He fell quite often and had many head injuries when he was young. He had an awkward gait and at 15 still has, walking with his feet inwards so he often trips. He also has enlarged 5th metatarsal bones on both feet, which makes getting the right shoes very difficult, as wide-fitting shoes are wide in the wrong place.

We were very lucky to have the most fantastic occupational therapist (OT), who started a "Hands" group for babies and children under 5 to get them to use their hands. The group was small, with only 5 or 6 children, some with a right hemiplegia. The children sat in a circle alongside their mothers and learned to place their hands flat on a ball, catch and pass it, as well as lots of other hand activities. This was great: it helped Colin, and I made friends with the other mothers which made me feel less alone. To start with, it was difficult to get Colin to sit still in the circle but the more he got used to going, the better he got. This OT made a difference to our lives with her work, patience and very caring manner, but sadly lost her battle with cancer a few years ago.

Professionals can make a huge difference to the life of their patients. I sometimes think if only they could step into my shoes, they might understand what it feels like to have a child with additional needs. You just need them on your side, helping you as best they can, instead of battling against them. Life is tough enough with a child with additional needs: professionals should be able to take a step back and think how they would want to be treated.

The geneticist explained Colin's disorder to us, and then we had a second visit to a different geneticist at our local hospital. Even though we had a firm diagnosis that our son had inv dup(15), this geneticist told us that he had Prader-Willi syndrome! As a parent, in my opinion geneticists do need to be absolutely certain that what they tell their patients is correct. If they are not absolutely sure they should never say: "Your child will not do this or do that." They should always give information on the relevant support organization at diagnosis if there is one available. Let the family decide whether they want to know more, and not assume that the professional knows what is best. Let the family decide whether they need support and when. This is better than leaving families to look up information on the internet and then find out more than they need to know.

When Colin started in the special needs assessment unit of the local specialist school at the age of 2, all his therapies were incorporated into the school setting, so that made life a little easier. From the age of 9 months Colin benefited from portage, a pre-school service for children under 5 with special needs. At first he received weekly visits from a support worker. By the age of 4, when he was more or less in school full time, he was receiving monthly visits. I cannot stress how important this service was to us as a family. It broke all tasks down into easy manageable steps, a practice I still use to this day.

Around the age of 2 Colin developed severe eczema. This was very difficult to manage for years: we tried all sorts of drugs and creams, lotions and wet wraps. At 6, he caught chickenpox and spent a week in hospital with chickenpox exacerbated eczema. This made him very ill: for the first 3 days he was on a drip and still has several white scars on his torso. At 7, after years of failed conventional treatments and frustration, our GP prescribed Colin gamolenic acid capsules (*evening primrose oil*). *This was to be a major breakthrough in clearing virtually all his eczema. To us it was like a miracle cure. Colin still gets occasional flare-ups in patches, but these are much easier to manage.*

When Colin was 18–24 months old he had some episodes of staring and dropping to the floor. Hospital investigations were inconclusive. Then, 3 days after his ninth birthday he developed absence seizures and now requires medication (*sodium valproate initially, with topiramate added a few years later*). *He has had occasional myoclonic seizures but these are rarer. At 16, Colin started having tonic clonic seizures, so the dosage of medications was increased.*

At 14 Colin choked on the skin of a sausage. At our local urgent treatment centre he quickly deteriorated and was taken by ambulance to another local hospital. He had aspirated and we were told his condition would deteriorate within the next 24 hours. He was given oxygen and monitored until a place was found at a hospital with pediatric intensive care, in order to perform a bronchoscopy. After a second chest X-ray, the pediatric consultant invited us to view the image and asked if we knew that Colin had a curve in his spine. This was news to us. He has a mild thoracic scoliosis at the top of the spine.

When he was 14, Colin had a video fluoroscopy because of my concerns about his swallowing. This confirmed that as he was given solids, smooth and liquid food and drink, he was aspirating on every mouthful. He was sent for immediate physiotherapy and following a discussion with the specialist speech and language therapist, we faced the decision of whether to let him have a gastrostomy. Instead of chewing his food, he rolls it around his mouth with his tongue and then appears to suck it down his throat. After much discussion and a home visit from the specialist speech therapist, we decided that we could manage his eating, especially as he normally likes to eat with his head tipped forwards, which lessens the risk of choking. We devised a plan of what he could and should not eat and how best to give it to him. The speech therapist visited both school and respite to make them aware of his needs and discuss the plan. We continue to manage this quite well except that I am still concerned he is aspirating, and am now awaiting another visit from the speech therapist with the echo voice microphone to check it out.

An MRI (*magnetic resonance imaging*) *when Colin was 6 years old showed an abnormality on the white matter. At some point, probably in utero, he had had a brain hemorrhage. My mind went straight back to his 20-week scan and the "black hole". I wondered whether this was what had showed up on the scan.*

Colin has severe dorsiflexion: his joints bend backwards and he has dislocated his hip and shoulders. His shoulders have occasionally "popped" out. He has always been hypotonic. He can stand on his feet with all of his toes bent underneath, and can stand facing one way with his feet facing in the opposite direction.

At age 16, Colin is unable to speak: he has no words. I assume this is either because of the brain hemorrhage, his hypotonia or his high arched palate. He is unable to use sign language, except for the sign for please/thank you but has had more success using PECS (*Picture Exchange Communication System*), *photos and objects of reference.*

Colin is able to walk, albeit in a slightly unusual uncoordinated fashion. He has no sense of fear or danger and has difficulty understanding different inclines or declines, so for example has to be told to step down off a kerb. We do use a wheelchair outdoors, as he can tire easily but is also very difficult to hold onto. He has great upper body strength and pulls hard at times.

Colin is unable to hold or write with a pencil or pen. He has very poor fine motor skills and does not like help to use his hands. He hates having his hands washed and if you try and get him to rub his hands together himself, as if to wash them, he will interlock his fingers.

Colin has always been fairly hyperactive and has been prescribed risperidone 0.5mg since about the age of 8. He is now much calmer. Some of his frustration could be due to his teenage years and not being able to make people understand his wants or needs. He does not have a malicious bone in his body, never bites, kicks or hits others and even if others hurt him, he does not retaliate. But he often bites his hand in frustration when he is trying to tell you something. He is a very loving boy. He very rarely cries so if he does, you know something is really wrong. He has a very high pain threshold. He has severe/profound learning difficulties and a sensory dysfunction (Fig. 6.20).

This year Colin moved from a special needs secondary school into the school's further education unit, which is like the sixth form in mainstream school. He likes the routine of school and loves swimming and horse riding.

As a parent, my worries now are what will happen to him when he leaves school at 19. The transition between child and adult services here is very poor and more work needs to be done to improve not only the transition process, but services for adults with learning disabilities in general.

As a baby Colin slept a lot: we had to wake him for feeds. As he got older his sleeping pattern changed. He would settle for sleep, but woke up in the night once, twice or sometimes many times. Aside from the same bedtime routine, we tried a variety of medications over the years including preparations of alimemazine, promethazine and chloral hydrate, none of which seemed to keep him asleep. At the time he was on no other medication. In later years melatonin was tried but that did not keep him asleep all night either. Today, Colin often still needs a nap during the day but now sleeps longer at night most of the time. His body clock appears to be upset by a change in moon cycles, when he will get up at 1 or 2 am and not go back to sleep, settle or relax. He appears to be worst affected about 2 days before a new or full moon.

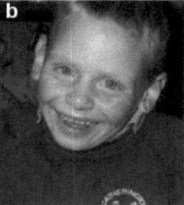

Fig. 6.20 Colin at the age of 9 months (**a**), 6 years (**b**), and 14 years (**c**) (copyright Unique)

6.16   Chromosome 16

6.16.1   Potentially Non-dose-sensitive Pericentric Region

The potentially non-dose-sensitive pericentric region of chromosome 16 spans at least 17.16 Mb, i.e., positions 28.86–46.02 Mb. Cytobands 16p11.2 to 16q12.1 are included therein (Fig. 6.21).

These data are supported by over 15 sSMC(16) cases and a number of familial cases with centromere-near duplications in the short arm and the long arm of chromosome 16 (Liehr 2011a). Also cryptic mosaicism has been reported; however, only in one case (case 16-O-p11.2~11.1/1-1; Liehr 2011a) this led to a mosaic partial tetrasomy in addition to the (mosaic) partial trisomy also present in the other cases.

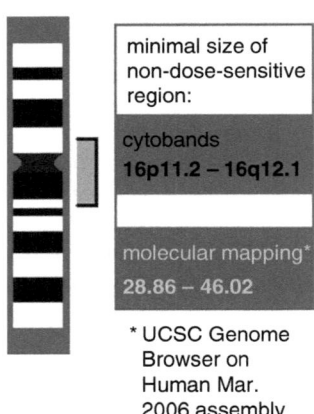

Fig. 6.21 Right of the ideogram of chromosome 16 drawn according to Kosyakova et al. (2009) the possibly non-dose-sensitive pericentric region is *highlighted*. The corresponding cytobands according to ISCN (*black bracket*) and the molecular mapping data (Mb) of the region are given (*light-gray square within black bracket*)

6.16.2   Clinical Signs

There are only six reports of sSMC(16) patients with clinical symptoms. The observed abnormalities are listed in Table 6.13. Dysmorphism and genital abnormalities were seen most frequently.

Table 6.13 Symptoms regularly observed in sSMC(16) cases with clinical symptoms (alphabetic order)

Symptoms	16 (%)
Dysmorphic face	83
Genital abnormalities	50
Hypotonia	17
Mental retardation	17

According to six informative cases collected in Liehr (2011a)

By contrast, intrachromosomal duplication of proximal pericentric regions of chromosome 16 was seen in over 15 cases with mental retardation, autism, growth retardation, and other features. However, the clinical significance of these imbalances is not clear, as at least 40% of them were inherited from one clinically normal parent (see case 16-W-IMB; Liehr 2011a).

6.16.2.1   Trisomy and Tetrasomy 16p or 16q

Partial trisomy of the long arm or of the short arm of chromosome 16 causes features similar to trisomy 18, being lethal during the prenatal or early postnatal period. Full trisomy 16 is not compatible with life (Schinzel 2001). Partial tetrasomy 16p or 16q has not been reported.

6.16.3   Mosaicism

Presently, it can only be stated that mosaicism is present in clinically normal and abnormal sSMC(16) carriers. Cryptic mosaicism is rarely seen.

6.16.4   Uniparental Disomy

More than 60 maternal UPD 16 and approximately ten paternal UPD 16 cases have been reported. In two cases maternal UPD 16 appeared together with an sSMC(16). For paternal UPD 16 there are three cases and for maternal UPD 16 there are two cases with activation of a recessive allele due to iUPD. It is commonly accepted that UPD 16 is not harmful because of imprinting effects. Nonetheless, it is a matter of discussion if the regularly observed intrauterine growth retardation in UPD 16 is the result of the mosaic trisomy 16 which is often detected together with UPD 16 in the placenta and/or the fetus (Schinzel 2001; Liehr 2011c).

6.16.5   Case Reports

Report A: See Sect. 4.2. Personal experience 3.
   Report B: Listed in Liehr (2011a) as 16-O-p11.2/2-1; reported by the mother of the now 3-year-old girl, with karyotype 47,XX,+r(16)(::p11.2->q12.2::)[8]/46, XX[9].

   Finding out I was pregnant was a happy, but scary proposition for us. Although I had had two easy and uneventful pregnancies and have two beautiful girls, who were born in 2000 and 2002, I suffered six miscarriages after that. Some of the causes of the miscarriages were understood (caused by various genetic abnormalities) and some were not. In addition, our doctors determined that I have a blood clotting disorder and elevated natural killer cells, both conditions that may also have caused or contributed to my miscarriages. The doctors concluded that for me to carry to term, I needed twice-daily injections of blood thinner and periodic IV infusions of immunoglobulin to control the natural killer cells.

Because of all the genetic abnormalities our pregnancies have had, we wanted to do early diagnostic testing with this pregnancy. If there was a genetic abnormality, we wanted to know so that we could really think about whether or not to continue with the pregnancy.

So, when I was 11 weeks pregnant, we went for an early amniocentesis, recommended by my obstetrician (OB). We have since learned that this procedure is a bit unusual, but the chorionic villi sampling (CVS) we had intended to do the week before could not be safely done because of the position of my placenta. A few days after the early amniocentesis, we got normal quick test results back. We were ecstatic, thinking that, finally, we had a chance of carrying a genetically normal fetus to term. Shortly after, however, we got the full genetic testing results back, that the fetus had mosaicism – nearly half of her cells contained an unidentified "marker chromosome," and further initial testing showed it was ring-shaped. Our online research revealed that this meant there was a very high chance that the baby would have cognitive and physical abnormalities – that sSMC in general were bad news and that the ring-shaped ones tend to be worse.

Our OB immediately arranged for the samples of the amniotic fluid to be sent to several different labs for more extensive testing. This is when we learned of Dr. Thomas Liehr and the very precise testing his lab can do. Dr. Liehr was sent amniotic fluid and eventually, fetal skin cells, too. We learned that identifying the exact chromosome the marker is from, and more specifically, what part of the chromosome the marker contains, could be the key to understanding what kind of symptoms the baby could have. Our doctor also sent us for a fetal skin biopsy. This test would determine whether the abnormal cells were actually in the fetus or were perhaps (hopefully) isolated to the placenta or amniotic fluid and therefore, would not affect the baby. The test itself was awful – painful and scary. The wait for results was even worse.

Needless to say, we were scared and concerned and began contemplating whether or not to continue with the pregnancy. Our OB was amazing – very thorough and very knowledgeable about these complicated genetic issues. We also recruited (through a cold call) a geneticist at Yale University who offered his help in deciphering the information we continued to receive. He was equally amazing, patient and comforting and helped us understand more about genetics than we ever wanted to know. In addition to the advice from our team of doctors, we spent our nights educating ourselves on the reality of outcomes given what we knew and did not know. We also reviewed Dr. Liehr's website and his extensive compilations of marker chromosomes and their clinical findings to further our understanding of reality.

A few weeks later, we finally received the results of the fetal skin biopsy. We were sad to learn that the abnormal cells were present in the fetus. The mosaicism percentage was less than 50%, so we were hopeful, despite almost no scientific evidence supporting this, that the "bad" cells were somehow dropping out. Each time we would get bad news, like the fetal skin biopsy results, we would also get news from Dr. Liehr and his team that maybe the marker had mostly heterochromatic (and therefore likely inactive) genetic material from around the centromere. Or, I would have an appointment for a sonogram, and at each appointment, the doctors would notice some new potential abnormality: odd-shaped kidneys, a lemon-shaped head, abnormal brain development, a stomach that was not filling and emptying properly. Then, I would go for a follow-up appointment with a specialist or another sonographer who would declare the organ development very typical and seemingly normal. So, we were back and forth, thinking that the baby would have some abnormalities, then thinking that perhaps the chances of abnormalities were fairly low.

Finally, Dr. Liehr identified the ring chromosome as follows: r(16)(::p11.21-> q12.2::). With this information, we were able to focus our research a bit more. We dug into the charts on his website, and learned what we could. It appeared that chromosome 16 markers can be asymptomatic. Through the Unique website, I found a woman in England who has a daughter with a somewhat similar, although smaller, marker. She and I became pen pals and friends. Her daughter has no clinical symptoms from her marker. Our geneticist at Yale University, Dr. Maurice Mahoney, eventually suggested that the chances of this baby having abnormalities were really no greater than that of a "normal" pregnancy.

All of this testing and researching and worrying and wondering and retesting and sonograms and appointments started when I was 11 weeks pregnant and resolved when I was nearly 30 weeks. At that point, we were nervous, but cautiously optimistic that the baby would be okay.

From then on, I went for twice-weekly non-stress tests and regular sonograms. At 35 weeks, I had a growth scan which measured the baby at 4 pounds, 15 ounces. Two weeks later, the scan showed that she was exactly the same size – no growth in 2 weeks. My doctor wanted me to get to the hospital to be induced as soon as possible. Plus, they thought they saw a stomach blockage on the sonogram. I knew there was not one (I had gotten pretty good at reading sonograms by this point!), but just to be safe, they sent me to be induced at a hospital where the surgeons on staff could operate immediately if need be. I was admitted in the evening, was induced the next day and delivered our baby girl (at 37.5 weeks gestation), who weighed exactly 4 pounds 15 ounces, had a very normal stomach and all of her fingers and toes. Her only physical abnormalities are two "dimples" or "sinuses," one near her tailbone and the other in the cartilage on one of her ears. She also has two scars on her legs from the fetal skin biopsy.

Interestingly, my cord blood was tested and the results showed a tiny percentage of abnormal cells with the chromosome 16 marker. My placenta, on the other hand, showed a very large percentage of full trisomy 16 cells.

So, our daughter is now 2 years and 5 months old and is a hilarious, sweet, adorable, and incredibly smart little girl. Her pediatrician has poked and prodded and examined and tested her, and has determined that she has no cognitive or physical limitations. We agree. She is extremely outgoing and social (way more than her big sisters were at this age), very verbal, very active and coordinated. She loves to talk to anyone she meets, she tells jokes, sings songs, loves her big sisters, and is a constant source of joy and laughter in our house. She makes us laugh every day and we are so grateful to have her. We think she is simply amazing.

6.17   Chromosome 17

6.17.1   Potentially Non-dose-sensitive Pericentric Region

In chromosome 17, cytobands 17p11.2 to 17q11.2 (very proximal part) seem to be potentially inert to gain of copy number. The region 18.68–23.32 Mb has been confirmed molecularly (Fig. 6.22).

However, these data are based on a single case and should be considered carefully. The three reported sSMC(17) patients without clinical symptoms do not show cryptic mosaicism or partial tetrasomy (Liehr 2011a).

Fig. 6.22 Right of the ideogram of chromosome 17 drawn according to Kosyakova et al. (2009) the possibly non-dose-sensitive pericentric region is *highlighted*. The corresponding cytobands according to ISCN (*black bracket*) and the molecular mapping data (Mb) of the region are given (*light-gray square within black bracket*)

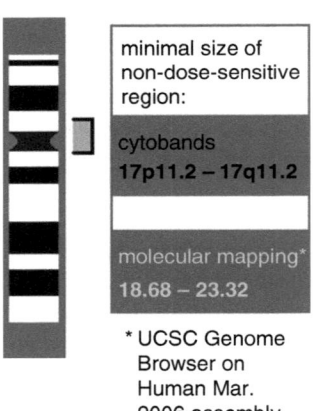

minimal size of non-dose-sensitive region:

cytobands
17p11.2 – 17q11.2

molecular mapping*
18.68 – 23.32

* UCSC Genome Browser on Human Mar. 2006 assembly

6.17.2   Clinical Signs

Proximal partial trisomy of chromosome 17 may in general lead to mental retardation, dysmorphism, and a number of more or less specific malformations (Table 6.14). Growth retardation is a more frequent feature in gain of copy number of the proximal part of the long arm of chromosome 17.

Table 6.14 Symptoms regularly observed in sSMC(17) cases with clinical symptoms (alphabetic order)

Symptoms	17p (%)	17q (%)
Clinodactyly	30	33
Dysmorphic face	60	67
Growth retardation	20	67
Heart defect	10	33
Hypotonia	50	33
Macrocephaly	30	–
Mental retardation	100	100
Microcephaly	20	67
Seizures	10	–

Cases distinguished by centromere-near partial trisomy of the short arm and the long arm according to ten and three informative cases, respectively, collected in Liehr (2011a)

Not listed in Table 6.14 is the fact that as soon as an sSMC(17) includes the region 17p12 (15.70–15.10 Mb, peripheral myelin protein 22, PMP22 gene), muscular neuropathy is the effect in the teenage or adult patient. As most corresponding case reports on sSMC(17p) cases were of small children, this is not mentioned there. However, threefold presence of PMP22 leads to Charcot–Marie–Tooth disease type 1A (OMIM #118220) or even Déjerine–Sottas syndrome (OMIM #145900).

6.17.2.1 Trisomy and Tetrasomy 17p

Two sSMC(17) leading to partial trisomy 17p (cases 17-W-p13.3/1-1 to 17-W-p13.3/2-1; Liehr 2011a) and more than ten patients with the same chromosomal imbalance due to other chromosomal rearrangements were noted (Schinzel 2001). Tetrasomy 17p is not known; however, trisomy 17p leads to intrauterine growth retardation, microcephaly with hydrocephalus, dysmorphic faces, male genital hypoplasia, clinodactyly, hypotonia, and severe mental retardation.

6.17.2.2 Trisomy and Tetrasomy 17q

Neither trisomy 17q nor tetrasomy 17q has ever been seen in a human fetus or newborn.

6.17.3 Mosaicism

Karyotypes 47,XN,+mar/46,XN are regularly observed in sSMC(17). Cryptic mosaicism and/or partial tetrasomy are, however, rare.

6.17.4   Uniparental Disomy

Fewer than five cases of maternal UPD 17, one case of paternal UPD 17, and fewer than five cases of UPD 17 of unknown parental origin have been reported (Liehr 2011c). sSMC(17) have not been reported with UPD 17.

6.17.5   Case Report

Provided by Unique; reported by the parents of a 16-year-old girl with karyotype 46,XY,dup(17)(p12p10). N.B.: Even though this patient has no sSMC(17) but a centromere-near duplication, the clinical outcome would have been similar to that for a patient with karyotype 47,XY,+r(17)(::p12->q11.1::).

Our daughter has a chromosome 17 duplication which involves the regions p10-p12. Unless you are a medical person, this means very little. In real terms it means she has a severe intellectual disability. She cannot talk, it took her 3 years to be able to sit up, she did not walk until she was 4½ years old and she needs help in every basic area of life: feeding, toileting, bathing, dressing and so on. What it does not tell you is that she is the most amazing person we know. Here is her story.

Dannielle [Note: most of the children's names have been changed in accordance with their parents' wishes] was born 16 years ago after a fairly uneventful pregnancy and very fast labor. The midwife told us that everything was fine, despite the fact that she did not cry when she was born, but there was a problem with her feet – she had talipes (clubfeet). It was very difficult to breastfeed her. Her temperature was low and so were her blood sugar levels, so she was taken to the newborn care nursery a few hours after birth. She was there for 3 days and was tube fed as she did not seem able to feed on her own. Then we were told all was well and that we could take her home. We only needed to follow up with the physiotherapist regarding her talipes. Her feet were placed in plaster and splints for many months and eventually at 16 months her right foot was operated on as it did not respond to splinting. In the meantime she was an exceptionally good baby who rarely cried and slept for most of the day and night. Feeding was a constant struggle and at 3 months we sought help to try and continue breastfeeding. The first thing the nurse did was put her finger in Dannielle's mouth and told us that she could barely suck. No wonder it was so difficult to feed her! All the while we were under the "care" of a pediatrician and had visited the baby health clinic for weekly check-ups. When I questioned her sleepiness I was told that she was probably still stunned from such a quick birth and that we were very lucky that we had such a good baby! Being the parents of a healthy, active 2-year-old we knew something was very wrong but it was very difficult to convince anyone in the medical profession of our concerns. We decided to bottle feed her. Again, no-one told us about squeeze bottles or any aid to help, so we struggled on. Most times it took one and a half hours to give her 50 ml of milk.

At about 7 months we tried our luck with a new pediatrician. She rolled for the first time on the morning of the appointment, so after the doctor had checked her over he felt that we essentially had a "normal" baby who was just slow to feed and reach her developmental milestones. She did not roll again for many, many months. She had been seeing a physiotherapist about her feet from birth and also for developmental delays but even they were not too concerned. When we questioned the fact that she could not sit without having pillows propping her up, we were told not to worry and to sit her in a box until she was strong enough to sit up of her own accord. Eventually she did sit up independently – the trouble was she was 3 years old!

We returned to the pediatrician for a follow-up appointment at 11 months of age and he saw a very different baby to the one at 7 months. He was very concerned about her lack of development and put her into hospital the next day for tests. He told us what the tests would be and that "one would be a chromosome test, but he doubted if that was the problem". So that word that has changed our lives so fundamentally was uttered for the first time. Sure enough, when the results of the blood tests came back, she did indeed have a chromosome abnormality – number 17 to be precise. We can never hear that number now and not think of our precious girl.

Like every parent of a disabled child, we have been on an incredible journey for the past 16 years. Nothing can prepare you for those words when you are sitting in a doctor's surgery and they tell you that your child has a severe intellectual disability and will probably never walk or talk or do anything that your other healthy child can do. The best you can hope for is that you can accept the news and do whatever it takes to help your child reach their potential. It might mean that they will never walk but they might be able to maneuver themselves somehow, they might never talk but they can communicate in their own way, they might never read or write or do anything that society regards as normal, but they will show you the true meaning of life and love. It will be heartbreaking at times and the grief will never leave you, but you will find a way to help them lead a happy and fulfilling life.

I am so glad we did not listen to all the "gloom and doom" from the doctors. Our child did walk. She was 41/2 years old, but she did it! She did it on partially dislocated hips that were not diagnosed by the medical profession (despite constant requests from us to X-ray). She did it through sheer determination and the fact that we could facilitate her desire to achieve this monumental goal. She had to use a standing frame to weight bear and a variety of walking frames to learn how to do it. She has had multiple orthopedic surgeries including her right foot, both hips and both knees. In addition and due to her chromosome 17 duplication, she was diagnosed with a peripheral neuropathy called Charcot Marie Tooth disease, which has contributed greatly to her orthopedic problems. She has had to endure hip dislocations and hip dysplasia and was operated on at 7 years of age and placed for 3 months in a "hip spica", which is essentially a full body cast from chest to ankles. She also had prolonged periods of kneecap dislocations before corrective surgery in 2007 and 2008.

OK, they got it right with her not talking but she can communicate basic needs in other ways. We use an augmentative communication program and she is learning

Fig. 6.23 Dannielle at the age of 4 months (**a**), 3 years (**b**), 6 years (**c**), and 14 years (**d**) (copyright Unique)

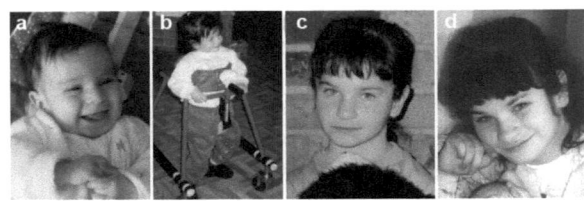

some signs. It is very difficult at times to work out what she wants but we usually get there in the end. We would probably say that this remains one of the biggest challenges in her life. The most difficult times are when she is sick and cannot say exactly what the problem is. It takes a good deal of guess work but you do learn over the years to "tune into" your child and most times you can work out what's happening. We would say the key to staying sane and "hanging in there" is to maintain your sense of humor and where possible, try and have a break from your child on occasions. Not only do you get the chance to recharge your batteries, but it does your child the world of good as well and gets them used to other people in their lives.

Toilet training has been a long and arduous process. We have been trying on and off for 11 years to get Dannielle toilet trained or toilet timed. Each time we would make some progress, surgery would put us back to square one but we still persist and she has made some significant gains in this area. Like many children with developmental disabilities, Dannielle's sleeping habits are difficult to cope with at times and a full night's sleep is a luxury. She has attended a special school for children with moderate to severe intellectual disabilities since just before she turned 4 years of age (Fig. 6.23).

She has made slow progress over the years in many areas. Her overall health is very good. She does not have any serious medical conditions such as epilepsy or heart problems. Dannielle is a very happy child who smiles readily and brings great joy to our family.

6.18   Chromosome 18

6.18.1   Potentially Non-dose-sensitive Pericentric Region

At a minimum the proximal pericentric region of chromosome 18, which is insensitive to gain of copy number, spans 12.80–17.30 Mb and/or cytobands 18p11.22 to 18q11.1 (Fig. 6.24).

If it is partial, proximal tetrasomy is tolerated only if it is present in a small percentage of the cells (cases 18-O-p11.22/1-1 and 18-Wi-158 and the mother of the patient in case 18-Wi-41; Liehr 2011a).

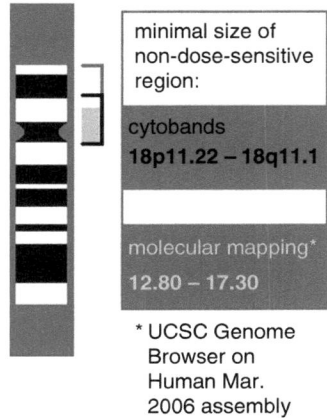

minimal size of
non-dose-sensitive
region:

cytobands
18p11.22 – 18q11.1

molecular mapping*
12.80 – 17.30

* UCSC Genome
Browser on
Human Mar.
2006 assembly

Fig. 6.24 Right of the ideogram of chromosome 18 drawn according to Kosyakova et al. (2009) the possibly non-dose-sensitive pericentric region is *highlighted*. The corresponding cytobands according to ISCN (*black bracket*) and the molecular mapping data (Mb) of the region are given (*light-gray square within black bracket*). The short arm is marked by a *gray half-bracket* as it may be exceptionally unproblematic if sSMC(18) is present in a mosaic

6.18.2 Clinical Signs

In general, sSMC(18) leading to symptoms, excluding those with isochromosome 18p (see Sects. 6.18.2.1 and 5.4), lead to rather nonspecific clinical signs (Table 6.15). Mental retardation and dysmorphic features can be observed, as are commonly found mostly in chromosomal imbalances.

Table 6.15 Symptoms regularly observed in sSMC(18) cases with clinical symptoms (alphabetic order)

Symptoms	18 (%)
Dysmorphic face	57
Heart defect	14
Mental retardation	57
Microcephaly	14

According to seven informative cases collected in Liehr (2011a)

6.18.2.1 Trisomy and Tetrasomy 18p

Trisomy of the whole short arm of chromosome 18 leads, only to very mild clinical signs and symptoms, if any, and normal to subnormal intelligence is reported (Schinzel 2001). The imbalance can be transmitted within families or can appear de novo. sSMC can be causative for the chromosomal imbalance, even though only complex rearranged ones have been reported for this condition (see cases 13/21-U27 and 13/21-U28; Liehr 2011a).

It is noteworthy that the 18p region can be tolerated more or less without problems in humans when present in three copies; however, four copies are associated with a severe phenotype – see Sect. 5.4 for i18pS.

6.18.2.2 Trisomy and Tetrasomy 18q

Tetrasomy 18q seems to be nonviable; however, trisomy 18q was reported as leading to features similar to those of full trisomy 18 (Schinzel 2001).

6.18.3 Mosaicism

Mosaicism can be observed in sSMC(18), whereas cryptic mosaicism is rather unusual. However, mosaicism has an influence on the clinical outcome, especially in cases with i18pS, as shown in Fig. 6.24 (see also Sect. 5.4).

6.18.4 Uniparental Disomy

Only one case with segmental maternal UPD 18 is known (Liehr 2011c).

6.18.5 Case Report

See Sect. 5.4.3.

6.19 Chromosome 19

6.19.1 Potentially Non-dose-sensitive Pericentric Region

In chromosome 19, cytobands 19p13.11 to 19q13.11 could be dose-insensitive; only the molecular region from 22.98–36.90 Mb has been confirmed (Fig. 6.25). In a mosaic form, even partial tetrasomy near the centromere is tolerated (case 19-O-p13.11/3-1; Liehr 2011a).

6.19.2 Clinical Signs

There are differences in the corresponding patients depending on whether the sSMC (19) is derived from the short arm or the long arm; however, mental retardation is seen in most patients with sSMC(19) (Table 6.16). In proximal partial trisomy 19q,

Fig. 6.25 Right of the ideogram of chromosome 19 drawn according to Kosyakova et al. (2009) the possibly non-dose-sensitive pericentric region is *highlighted*. The corresponding cytobands according to ISCN (*black bracket*) and the molecular mapping data (Mb) of the region are given (*light-gray square within black bracket*)

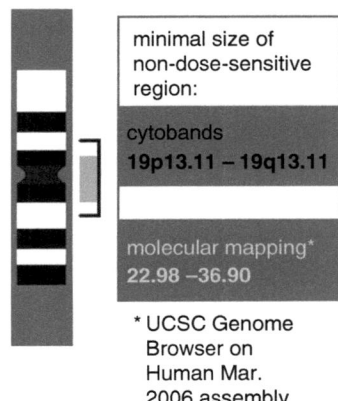

minimal size of non-dose-sensitive region:

cytobands
19p13.11 – 19q13.11

molecular mapping*
22.98 –36.90

* UCSC Genome Browser on Human Mar. 2006 assembly

Table 6.16 Symptoms regularly observed in sSMC(19) cases with clinical symptoms (alphabetic order)

Symptoms	19p (%)	19q (%)
Clinodactyly	40	–
Dysmorphic face	40	57
Growth retardation	40	14
Heart defect	–	14
Hypotonia	40	14
Macrocephaly	–	28
Mental retardation	80	100
Microcephaly	20	–
Seizures	60	–

Cases distinguished by centromere-near partial trisomy of the short arm and the long arm according to five and seven informative cases, respectively, collected in Liehr (2011a)

seizures, clinodactyly, growth retardation, hypotonia, and microcephaly are present more frequently and/or exclusively. More frequently expressed in partial trisomy 19q cases are dysmorphic face, macrocephaly, and heart defects.

6.19.2.1 Trisomy and Tetrasomy 19p

Mosaic trisomy 19p was seen in connection with an sSMC(19), leading to multiple dysmorphic signs at 1 year of age (case19-CW-4; Liehr 2011a). It could be that individuals with nonmosaic trisomy as tetrasomy 19p are not viable.

6.19.2.2 Trisomy and Tetrasomy 19q

Babić et al. (2007) reported the first case of trisomy of the entire long arm of chromosome 19; it was recognized prenatally owing to massive sonographic findings. Tetrasomy 19q has not been seen.

6.19.3   Mosaicism

Cryptic as well as cytogenetically visible mosaicism can be present in sSMC(19) cases with and without clinical signs. No conclusions can be drawn in this connection.

6.19.4   Uniparental Disomy

Only one case with segmental UPD 19 of unknown parental origin is known (Liehr 2011c).

6.19.5   Case Report

Provided by Unique; reported by the parents of an 11-year-old girl with an imbalance in chromosome 19 characterized by aCGH as arr 19p13.3(1372163-4368983)x3,19p13.3-p13.2(6802818-10978420)x3. As no cytogenetic result is available, the imbalance can be either due to an sSMC(19) or a duplication event.

Rose [Note: most of the children's names have been changed in accordance with their parents' wishes] is our first adopted daughter. I remember when we first brought her home; she was always smiling and loved to be around us. She could easily be entertained with the local news or a football game; she could sit for hours watching. When Rose was just a few months old, I remember taking her to her first genetics appointment with a well known doctor who could be seen on our local news. After examining Rose for about 15 minutes he looked to my husband and me and said so matter-of-factly: "Rose will never walk or talk and will not be a normal 21-year-old." I was mortified! How could he possibly predict her future? We never saw him again! Rose is 10 years old now and walks and has some single word utterances, not always clear or understood by those who do not know her, but nonetheless she is talking.

Rose could not meet the typical milestones that infants usually make. Rose had bilateral hip dysplasia at birth and underwent hip surgery at several weeks old. After surgery she was in a body cast for 12 weeks so she could not roll, sit up or crawl. So her milestones were quite delayed. When she was about 4 years old, Rose had not as yet drunk from a straw. We were in a restaurant having lunch and I had walked away to get some napkins. Well, she wanted my soda I guess in a bad way and picked up my drink and sipped through the straw. I was so excited I was yelling in the crowded restaurant "Did you see that? Rose drank through the straw!" Everyone in the restaurant was looking at me like I was off my rocker!

Rose is a pretty medically involved child. She has 15 doctors she sees on a regular basis. Some of Rose's diagnoses are: visually and hearing impaired; orthopaedic issues; developmentally delayed; periodic syndrome; Hashimoto's

Fig. 6.26 Rose at the age of 9
years (copyright Unique)

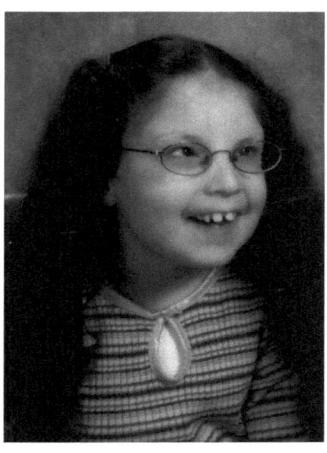

thyroiditis; pervasive developmental delay; pica and hypogammaglobulinaemia just to name a few. To look at Rose, she looks like a typical child, always happy and always has a smile, with a great sense of humor. She can draw your attention and grab hold of your heart for a very long time. Rose has a sister Kay [Note: most of the children's names have been changed in accordance with their parents' wishes] who also has a chromosome imbalance (16q duplication) and is 3 years younger. They are both now at a stage where they are very interactive with each other. Rose is very protective of Kay and is very troubled if Kay is crying. Rose will come get us and continue to say "Kayeee, Kayee, what" (Fig. 6.26).

Both Rose and Kay are adopted and are not biologically related. When we adopted Rose we knew from the start that she would probably have some significant issues. After a few years we decided that Rose needed someone to relate to so she would not feel different. So we adopted Kay. It was the best decision for both children. They are both thriving and doctors are amazed at how far both girls have come. They would never have predicted they would be doing so well!

6.19.6   Chromosomes 1/5/19

See Sect. 6.1.6.

6.20   Chromosome 20

6.20.1   Potentially Non-dose-sensitive Pericentric Region

The putatively non-dose-sensitive pericentric region of chromosome 20 includes at least molecular positions 24.96–29.93 Mb and cytobands 20.p11.22–20.q11.21 (Fig. 6.27).

Fig. 6.27 Right of the ideogram of chromosome 20 drawn according to Kosyakova et al. (2009) the possibly non-dose-sensitive pericentric region is *highlighted*. The corresponding cytobands according to ISCN (*black bracket*) and the molecular mapping data (Mb) of the region are given (*light-gray square within black bracket*)

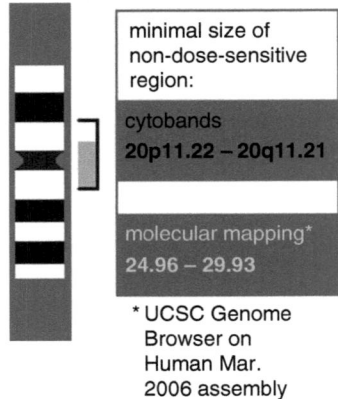

Cryptic and noncryptic mosaicism in connection with proximal partial tetrasomy can also be tolerated (cases 20-O-p12.2/1-1, 20-O-p11.21~11.22/1-1, 20-O-p11.21/1-1, and 20-O-p11.1/3-1; Liehr 2011a).

6.20.2 Clinical Signs

Clinically, sSMC(20) derived from the long arm or the short arm cannot be distinguished (Table 6.17). Clinical signs are also more or less nonspecific.

Table 6.17 Symptoms regularly observed in sSMC(20) cases with clinical symptoms (alphabetic order)

Symptoms	20p (%)	20q (%)
Clinodactyly	27	20
Dysmorphic face	83	60
Growth retardation	17	20
Heart defect	50	20
Mental retardation	67	60

Cases distinguished by centromere-near partial trisomy of the short arm and the long arm according to six and five informative cases, respectively, collected in Liehr (2011a)

6.20.2.1 Trisomy and Tetrasomy 20p

Partial trisomy 20p has a distinctive phenotype, according to Schinzel (2001). One of these cases was due to a neocentric sSMC(20) – see Sect. 7.20. Growth retardation, dysmorphic faces, coarse hair, clinodactyly, and heart defects were found. Tetrasomy 20p was only seen twice, leading to severe malformations (Wu et al. 2003a; Fryer et al. 2005).

6.20.2.2 Trisomy and Tetrasomy 20q

A few cases have been reported for trisomy 20q (Sepulveda and Be 2008), a condition which leads to massive physical malformations. Mosaic isochromosome 20q has been reported prenatally in connection with monsomy 20p and a karyotype of 46 chromosomes. In most cases this karyotype is limited to the fetal bladder cells; the latter make a large contribution to the cells studied in amniocentesis (Dupont et al. 1997). In rare cases this karyotype can also be present in all body cells, leading to sonographic abnormalities (Goumy et al. 2005).

6.20.3 *Mosaicism*

Cryptic mosaicism has been reported only for clinically normal sSMC(20) carriers; cytogenetically visible mosaicism can but does not have to be present in clinically abnormal carriers.

6.20.4 *Uniparental Disomy*

Two cases of maternal UPD and one case of paternal UPD associated with sSMC (20) are known. In addition, over 15 more UPD 20 cases have been reported. The clinical significance of maternal UPD 20 is under discussion. However, parts of chromosome 20 are subject to imprinting, as paternal UPD 20 was recently correlated with pseudohypoparathyroidism (OMIM #103580, #603233, #612462) (Liehr 2011a).

6.20.5 *Case Report*

Provided by Unique; reported by the parents of a 4-year-old girl with karyotype, 47, XX,dup(20)(p11.22q11.21),+r(20).ish dup(20)(p11.22q11.21),+r(20) oligo arr cgh dup(20)(p12.2q12)[11]/46,XX,dup(20)(p11.22q11.21).ish dup(20)(p11.22q11.21) oligo arr cgh dup(20)(p12.2q12)[8]/46,XX.

After a completely normal pregnancy, Ava [Note: most of the children's names have been changed in accordance with their parents' wishes] blessed us by taking her first breath in 2007. I had not any extra testing during the pregnancy because it would never change the outcome of any of my pregnancies; thus I saw no need to. All of my ultrasounds and measurements were perfect. We already had a very healthy 18 month old son at the time of Ava's birth so we planned on being very busy, but we had no idea how busy at the time. We thought that everything would be as perfect as the delivery with my son, and in some ways it was, but with a few "extra bumps in the road". Ava was about to show us a new level of thinking and a new level of LOVE.

The day after Ava was born, the delivery nurse thought that she had some characteristics of Down syndrome and that we needed to do genetic testing on her. She had upslanting eyes (which looked more pronounced right after delivery because of swelling), bilateral simian crease on her hands, tiny ears, and a sandal gap between her toes. They were all very mild characteristics so the healthcare professionals were not sure. Her muscle tone and reflexes were perfect at birth, her APGAR was 9, and she was otherwise a healthy baby. The pediatrician was having a hard time believing that she had Down syndrome, but thought that genetic testing would put all questions to rest. We went home with her the very next day having felt like we were hit with a ton of bricks.

We waited 2 long weeks for the results of the tests. The phone call finally came and we were told that Ava did not have Down syndrome, but she did have a very rare chromosome abnormality! She had an abnormality on chromosome 20. She had some cells with normal chromosomes, some cells with duplication on the centromere of chromosome 20, and some cells with an extra supernumerary ring made up of the same material as the duplication on chromosome 20. Her karyotype is a mile long. It is very hard to even explain it to most people and even some doctors. My husband and I had our chromosomes tested as well, and our blood work showed normal chromosomes. They said that Ava's abnormality was de novo, meaning that we were not carriers of anything, that it just happened during cell division for some reason.

Ava was born with two holes in her heart that have closed on their own already. She has urinary reflux which is treated with prophylactics currently, but we are hoping to have the surgery to correct it soon. She has sleep apnea and had her tonsils and adenoids out at 1 year of age. The surgery helped tremendously but she still has central sleep apnea which is believed to be due to a diagnosis of Chiari malformation Type I. She has Chiari malformation Type I, but currently has no symptoms from it other than the sleep apnea. She had tubes put in her ears at age 2 and had a vocabulary explosion after the tubes were placed! She is hitting all of her milestones on the late end of normal development so far, and we pray that she continues to do well in the future. She has been in therapy since birth, and it is a big help for all of us. She sat up and crawled at 8 months, walked on her own at 14 months, and today, at age 4, is running everywhere that you can imagine (Fig. 6.28).

The only therapy she is in now is speech, and she is making improvements in leaps and bounds in talking today. We used sign language a lot from age 2–3, and then she dropped most of the signs and replaced them with words. She is talking in complete sentences now and we are working on articulation in speech. Some things are still hard to pronounce for her, but most things are understandable to everyone. She amazes us every day and we are so proud of her achievements thus far. She is growing at a normal height and weight for her age, and is developing into such a beautiful little girl. She is completely potty trained now. She is very

*friendly and says hi to everyone in the stores. She loves all of her family members
and gets the biggest smile when family comes to visit. She loves playing with other
children and especially her big brother. She plays well with other children and has
the normal struggles of not wanting to share her toys. Her favorite toys are baby
dolls, her Barbie car, and Wubbzy. She loves to be outside and enjoys playing on
her swing set. She loves to ride her 3-wheel bike everywhere. She does not sit still
for very long and keeps us very busy. She loves to watch the cartoons Wubbzy,
Dora, and Alvin and the Chipmunks. She loves going on trips in the car, and we
plan on traveling a lot with both of our children. We live very normal lives but with
a few extra worries here and there. We do not dwell on the worries today. We take
each trial as it comes. We look forward to watching her grow into a beautiful young
girl and hope many things for her future, but we will take it day by day and continue
to love her for who she is.*

*   The two biggest things that Ava has taught us are that it is easier to live day by
day instead of worrying about the future, and she has taught us what complete trust
in God is. We worried so much in the beginning, but we finally let the worry go and
gave it to God. Ava has changed our family in so many ways and we would not
change one thing that got us to where we are today. We are so thankful for the
chance to raise beautiful children, and we are so thankful that Ava showed us the
real meaning of living a happy life. We all hope for perfect children and we have
come to learn that – no matter what – all of our children are beautifully and
wonderfully made, and perfect in their own way!*

6.21   Chromosome 21

6.21.1   Potentially Non-dose-sensitive Pericentric Region

In chromosome 21, another acrocentric chromosome like chromosomes 13–15,
the entire short arm is non-dose-sensitive. Furthermore, a small part of the proximal
long arm is dose-insensitive, i.e., cytobands 21pter to 21q21.1 and at least molecularly 0.00–14.85 Mb (Fig. 6.29).

Fig. 6.29 Right of the ideogram of chromosome 21 drawn according to Kosyakova et al. (2009) the possibly non-dose-sensitive pericentric region is *highlighted*. The corresponding cytobands according to ISCN (*black bracket*) and the molecular mapping data (Mb) of the region are given (*light-gray square within black bracket*)

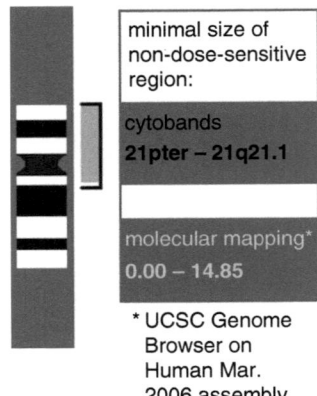

minimal size of non-dose-sensitive region:

cytobands
21pter – 21q21.1

molecular mapping*
0.00 – 14.85

* UCSC Genome Browser on Human Mar. 2006 assembly

Most of the clinically normal patents with sSMC(21) are nonmosaic, and only one patent has a cryptic mosaicism leading to partial tetrasomy (case 21-O-q11.21/1-1; Liehr 2011a).

6.21.2   Clinical Signs

In sSMC(21) with clinical signs one has to distinguish between sSMC(21) with and without the Down syndrome critical region. The latter normally show patterns similar or identical to Down syndrome with three copies of a whole chromosome 21.

In the five clinically abnormal patients with sSMC(21) excluding the Down syndrome critical region (Table 6.18), the clinical signs are nonspecific. The possibility of a cryptic, undetected mosaic of 47,XN,+21/47,XN,+mar(21)/46, XN cannot be excluded (Stefanou and Crocker 2004).

Table 6.18 Symptoms regularly observed in sSMC(21) cases with clinical symptoms (alphabetic order)

Symptoms	21q (without Down syndrome critical region) (%)
Autism	20
Dysmorphic face	100
Growth retardation	60
Hypotonia	20
Mental retardation	100
Microcephaly	20

According to five informative cases collected in Liehr (2011a)

6.21.2.1   Trisomy and Tetrasomy 21q

Trisomy 21 leads to Down syndrome (OMIM #190685). Tetrasomy 21 is lethal (Liehr et al. 2001).

6.21.3 Mosaicism

Cytogenetically visible and cryptic mosaicism may occur in sSMC(21). However, partial tetrasomies are seen rather seldom (Liehr 2011a).

6.21.4 Uniparental Disomy

Overall, 12 UPD 21 cases have been described. In seven cases UPD 21 was maternally derived, in four cases UPD 21 was paternally derived, and in one case UPD 21 was of unknown origin. An sSMC has not been associated with UPD 21; also no activation of a recessive allele has been reported (Liehr 2011c).

6.21.5 Case Reports

Report A : Listed in Liehr (2011a) as 21-W-q11.2~21.1/1-2; reported by the mother of the now 10-year-old boy, with karyotype 47,XY,+min(21)(pter->q11.2~21.1:) [5]/46,XY[10].

Our son Zaki [Note: most of the children's names have been changed in accordance with their parents' wishes] was born at the hospital in 2001. He was pronounced fit and healthy and we went home with our new bundle. For the next year everything was fine except that I had noticed that he was behind with all his developmental milestones. He did not crawl until he was 1; he sat up at 10 months and did not walk till 19 months. We put it down to the fact that he was a very heavy chap as we had been told to "stuff him up" at birth because he was very small.

He had some physiotherapy and that seemed to get him moving and mobile. He had and still does the happiest disposition of any child I know. I started to take him to playgroups so as a way of meeting other kids and for me to meet some other mums. I noticed much more the difference with him to the other kids. He was not interested in playing with them; he just wanted to do his own thing, usually climbing the stairs! He seemed to lack dexterity as well and would avoid things where he would have to try too hard.

In January 2004 my health visitor expressed concerns about him regarding his lack of speech and ability to do the same things as his peers. At the time I was not too bothered thinking that all kids do things in their own time. However, more unusual behavior had started to appear. He had a fascination with heat and would put his hand on the radiator and it never seemed to burn him. He would make odd sounds at other children but not adults and he seemed to have a heightened sense of smell. Sometimes he would stop and smell people's feet in the street! He was incredibly strong and appeared to have a very good memory of places and people he had seen before.

As time progressed his behavior worsened making just venturing to the shops or getting on a bus a major ordeal. People would make comments that were very hurtful and considering they knew nothing of my situation, very unjust. This still happens to this day and is the one thing that I have not managed to overcome myself.

In summer 2004 I went with my son to go and see the nursery he was starting in autumn. I had told them my concerns about his behavior but until that day I do not think they realized how bad it was. When I took him in he did not know what to do with himself. He just threw the toys, tried to hurt the other kids and wanted to escape. At that point one of the other parents came up and asked me if he was autistic. I sat and cried in the nursery. I had thought this but on speaking with my mum and his dad they thought I was overreacting as they never saw him behave like this. I knew at this point I was right.

He started nursery but was only allowed to go for an hour a day which eventually became 2 after a month. I requested a statement for him so he could get proper help. So started a very long process of specialist support services, educational psychologist, speech therapy etc. Looking back, nursery did not have a clue how to support a child like my son and just kept saying at least he can come for 2 h. All the other kids were there for 6 h and because of his disability he was excluded full-time and went to an Inclusion Childminding Service which he really did enjoy. Alas, this was only for 10 weeks. All my plans to go back to work were out of the window as there was no-one to look after him and I really felt alone in my battle for his statement. I had to go to my main profession, research stuff myself; put forward ways of trying to help and the wait of 6 months is far too long. Zaki never did go to nursery fulltime. They would only take him for his statemented hours of 20 h a week.

In April 2005 at my request my son had blood taken to test for any genetic abnormalities. I wanted to rule out everything for my son's weird and wonderful behavior. He was so aggressive now. He would hit kids as they walked past in the street, have major tantrums on the bus or at the shops. I just could not understand why. The blood test results came back in May but our doctor never told us till July! We went to see him and he said he would ask them to contact us with what it was as it was impossible to decipher unless you were an expert. He also failed to read the report properly as it asked for myself and his dad's blood to be sent to them to see if we were carriers for this mystery condition (Fig. 6.30).

Finally in August we got a copy of this report and read what I had always felt in my guts. Zaki had been diagnosed with a very rare genetic disorder called "mosaic trisomy 21" (without Down syndrome critical region). I threw myself into the computer researching this problem, contacting various organizations for help and just trying to make myself more aware of what it was. In October we had our blood tests and finally met with the clinical geneticist who explained we were not carriers and it was an unfortunate incident at conception. Still I sit here and think "why me"?

Zaki is now at school fulltime and making excellent progress. He will always be "special" and have his quirks about him but he is learning and surrounded by lots of lovely school friends who are becoming more understanding of his ways.

Fig. 6.30 Zaki at the age of
8 years and his mother

As for the future, only time will tell. I hope to see more research done into children with behavioral and learning difficulties. Medical practitioners need all the help they can get as relatively little is known about these disorders and their symptoms and too many kids are written off as naughty or given an incorrect diagnosis and the parents are left wringing their hands in despair at the lack of information.

Report B : *Provided by Unique; reported by the parents of a boy with karyotype 46,XY,trp(21)(q11.2q22.1). N.B.: Even though this patient has no sSMC(21) but a triplication near the centromere, the clinical outcome would have been similar to that for a patient with karyotype 48,XY,+del(q22.1)x2.*

Joe [Note: most of the children's names have been changed in accordance with their parents' wishes] was born 4 weeks early and everything seemed fine. After a few weeks he started screaming for nothing and it was high-pitched and could go on for hours. Nothing could be done to stop it.

When he was about 4 months old he was admitted to hospital after going floppy and not responding to anything. Tests were done and nothing found and he was allowed home after a few days. At about 6 months old, we had a follow-up appointment with the consultant who noticed that his development was slow and he was not sitting very well. We mentioned some of the other problems we were having, such as drinking, sleeping and screaming. More tests were done and it was found that Joe had duplication on chromosome 21. The name we were given was interstitial duplication of one long arm of chromosome 21.

We were given an appointment with a geneticist. She told us it was not Down syndrome as Joe only had part of an extra chromosome, but there could be similarities due to it being number 21. We were told there could be learning, physical and behavior problems but it was impossible to say how severe. We would have to wait and see what happened.

Joe's physical development was always slow but at 9 years old he can now do most things such as walking, running and jumping. Learning has been the biggest problem and Joe still cannot talk, but he uses Makaton signs and attends a special needs school. He is still not aware of danger and needs constant supervision.

Behavior has always been challenging and he can throw tantrums for nothing. We had some tests done for attention deficit hyperactivity disorder and he was said to be borderline but nothing else was ever said about it. I and Joe's father had some blood tests done to see if the problem had come from one of us but our chromosomes were OK. We were told that if we had any more children there was a very slim chance of it happening again.

When Joe was 5 years old, our twins Kylie and Callum [Note: most of the children's names have been changed in accordance with their parents' wishes] were born 8 weeks early. They had to stay in the neonatal unit for 4 weeks and this is when Callum started to scream so we decided to have him tested. The results came back and he had the same condition as Joe but we were now told it was a not a duplication but a triplication. We have since been told that Joe's is a triplication as well. Callum is small like Joe and is like him in many ways but he seems to be making more progress. At the age of 3, Callum attends a mainstream nursery but we are unsure whether he will cope in a mainstream school as he still needs constant supervision and help. Callum takes sodium valproate for seizures and Joe takes melatonin to help him sleep. We have looked on the internet for information about their chromosomes but can only ever find facts about Down syndrome (Fig. 6.31).

Kylie has since been tested and is OK so it may only affect boys in our family. We have now been told that if we have any more boys they would most likely have the same condition. We have been on the Unique website and have not found anyone who can help us yet but both boys are making slow progress.

Fig. 6.31 Callum and Joe (copyright Unique)

6.21.6   Chromosomes 13/21

See Sect. 6.13.5.

6.22 Chromosome 22

6.22.1 Potentially Non-dose-sensitive Pericentric Region

sSMC derived from the acrocentric chromosome 22 have a non-dose-sensitive short arm. Additionally, cytoband 22q11.21 is dose-insensitive, correlating to molecular level 0.00–16.37 Mb (Fig. 6.32).

If larger segments of 22q11.21 are included, an abnormal clinical outcome can be expected. However, as shown in Fig. 6.32 by the gray half-bracket, mosaicism can modulate clinical consequences, as in cases 22-O-q11.21/1-1 to 22-O-q11.21/5-1 (Liehr 2011a).

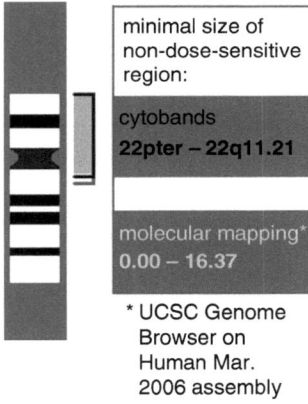

Fig. 6.32 Right of the ideogram of chromosome 22 drawn according to Kosyakova et al. (2009) the possibly non-dose-sensitive pericentric region is *highlighted*. The corresponding cytobands according to ISCN (*black bracket*) and the molecular mapping data (Mb) of the region are given (*light-gray square within black bracket*). The small part marked by a *gray half-bracket* highlights a region which may be unproblematic if sSMC(22) is present in a mosaic

6.22.2 Clinical Signs

Clinically, proximal partial trisomy 22 can contribute to ES (see Sect. 5.1), and proximal partial tetrasomy 22 can contribute to CES (see Sect. 5.2).

In other rare instances, partial trisomy 22 can lead to clinical signs and symptoms such as those listed in Table 6.19 for nine cases, which are relatively nonspecific.

Table 6.19 Symptoms regularly observed in sSMC(22) cases with clinical symptoms (alphabetic order)

Symptoms	22q (%)
Dysmorphic face	67
Duane syndrome	22
Heart defect	22
Hypotonia	56
Mental retardation	89

According to nine informative cases collected in Liehr (2011a)

6.22.2.1   Trisomy and Tetrasomy 22q

Trisomy 22, also due to 46,XN,i(22q) karyotypes, is known as a prenatally lethal condition (Schinzel 2001). However, mosaics can survive the perinatal period as well (Hall et al. 2009). Tetrasomy 22q has not been seen.

6.22.3   Mosaicism

Mosaicism can moderate or modulate clinical symptoms in the case of mosaic sSMC(22) normally leading to CES (cases 22-O-q11.21/1-1 to 22-O-q11.21/5-1; Liehr 2011a). Otherwise, the influence of (cryptic) mosaicism remains to be elucidated.

6.22.4   Uniparental Disomy

Maternal UPD 22 has been reported in 12 cases and paternal UPD 22 has been reported in fewer than five cases. Two sSMC(22) cases with maternal UPD 22 and one case with paternal UPD 22 are among them. One case was a clinically normal patient, one case was a CES patient, and one case was an ES patient. Neither imprinting nor activation of a recessive allele has been reported for chromosome 22 (Liehr 2011c).

6.22.5   Case Reports

See Sects. 5.1.3 and 5.2.3.

6.22.6   Chromosomes 14/22

See Sect. 6.14.6.

6.23   X Chromosome

6.23.1   Potentially Non-dose-sensitive Pericentric Region

For the X chromosome in cases with karyotype 47,XN,+mar(X), no dose-insensitive region has been defined apart from the centromere, matching cytobands Xp11.1 to Xq11.1 or 56.60–65.00 Mb (Fig. 6.33) (Liehr 2011a).

Fig. 6.33 Right of the ideogram of the X chromosome drawn according to Kosyakova et al. (2009) the possibly non-dose-sensitive pericentric region is *highlighted*. The corresponding cytobands according to ISCN (*black bracket*) and the molecular mapping data (Mb) of the region are given (*light-gray square within black bracket*)

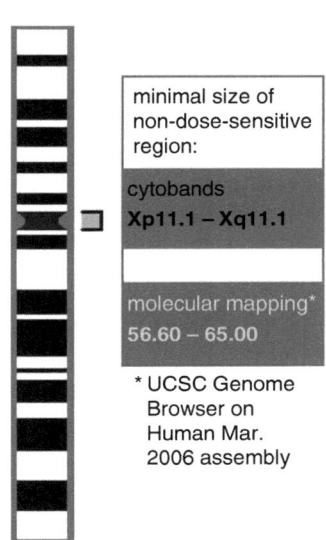

minimal size of non-dose-sensitive region:

cytobands
Xp11.1 – Xq11.1

molecular mapping*
56.60 – 65.00

* UCSC Genome Browser on Human Mar. 2006 assembly

6.23.2   Clinical Signs

In patients with clinical signs, sSMC(X) normally lack the X-inactivation (XIST) region. Thus, if euchromatin is present on the sSMC(X) and the XIST region is absent, corresponding genes are expressed as well as those on the normal (active) X chromosome (see also Fig. 5.5). Thus, such sSMC(X) with clinical signs are correlated with mental retardation. Other observed features are listed in Table 6.20; however, they are more or less nonspecific, even though obesity was present in 40% of these cases.

If an additional derivative of the X chromosome including the XIST region is present, the corresponding carrier is normally not (significantly) mentally or otherwise impaired (Schinzel 2001).

6.23.2.1   Trisomy and Tetrasomy Xp

Neither partial trisomy Xp nor partial tetrasomy Xp has been reported.

Symptoms	X (%)
Table 6.20 Symptoms regularly observed in sSMC(X) cases with clinical symptoms (alphabetic order)	
Clinodactyly	40
Dysmorphic face	60
Growth retardation	10
Hypotonia	20
Mental retardation	100
Microcephaly	20
Obesity	40

According to ten informative cases collected in Liehr (2011a)

6.23.2.2   Trisomy and Tetrasomy Xq

Trisomy or tetrasomy of the long arm of the X chromosome should be without severe clinical problems, as the XIST region should ensure the corresponding isochromosome or derivative X chromosome is inactivated (Höckner et al. 2009). However, the X-inactivation can also lead to the activation of an X-chromosomal recessive gene mutation in females (Ou et al. 2010).

6.23.3   Mosaicism

Both types of mosaicism are observable in sSMC(X) cases: cytogenetically visible and cryptic ones. No clinical correlation can be made with respect to mosaicism.

6.23.4   Uniparental Disomy

UPD X has been not seen in females with sSMC(X). For further details on the complicated issue of UPD X, see Liehr (2011c).

6.23.5   Case Report

Provided by Unique; reported by the mother of a 5-year-old boy with karyotype 47, XY,+r(X)(::p11.2->q12::)[39]/46,XY[11].ish r(X)(DXZ1+,XIST-).

Thomas [Note: Most of the children's names have been changed in accordance with their parents' wishes] was born in 2006. It was a long labor and there was meconium present, meaning the baby was distressed. He weighed 5lb 12oz (2.6 kg), quite small compared to my other two children who were both 8lb (4 kg). He was slow to feed, so needed special care in the baby unit for 4 days.

At home he never really got into a routine with his feeding and cried a lot. At 6 weeks he got bronchitis and was in hospital for 2 weeks. He then got a bad eye infection due to a blocked tear duct and had to undergo a surgical procedure. He then picked up a rotavirus infection and was back in hospital for Christmas in 2006. The doctors said he had a very low immune system.

At home Thomas was getting sick and suffering with chest infections. At 10 months he was not thriving, he was very floppy and was unable to sit up, and I knew as a mother that something was not right.

We got an appointment to see a consultant pediatric neurologist. Some blood tests were carried out, and the results showed that Thomas had an extra piece of chromosome in ring form attached to his X chromosome which was one of his sex chromosomes. The professor indicated that this was very rare and was the cause of his developmental delay. This news was a lot for myself and his father to take in, as we did not know what lay ahead for us.

At 18 months we got Thomas into Early Services. At this stage he was making no attempt to crawl or to talk and we were getting very frustrated. In order to get around, he would just roll but eventually, with the help of physiotherapy, he took his first steps around the Christmas of 2009. Now, at 3½ years, he is running around and loves to be outside.

Thomas is now also starting to understand us better. He will bring us things upon request. We are still a bit concerned about his speech, as the only word he uses sounds like "Dad", which he uses to refer to everything.

Thomas attends a special pre-school five mornings per week where he receives physiotherapy, speech and language and occupational therapy (Fig. 6.34).

Thomas is a lovable child. He has huge blue eyes and eyelashes you would die for. His hair is very coarse and thick and grows very fast. He hates getting his hair cut – it takes two of us to hold him. He loves his "Balamory" DVDs and loves toys that play music. He is very strong willed and tries to bite us sometimes when he does not get his own way. He loves being with our extended family and going for walks. We all love and adore him.

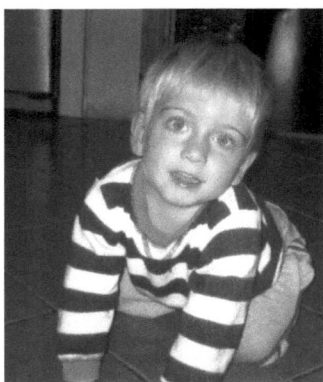

Fig. 6.34 Thomas (copyright Unique)

6.24   Y Chromosome

The Y chromosome is present in only one copy in males and has (in general) no really adverse influence if it is present twice in so-called XYY syndrome. Thus, an sSMC(Y) may cause some trouble in sex determination; otherwise it should be inert. However, recently a slightly decreased lifespan has been discussed for XYY carriers (Stochholm et al. 2010).

6.24.1   Case Report

See Sect. 5.5.3.

6.25   Centric sSMC of Unknown Origin

As well as the well-characterized centric sSMC discussed before (see Sects. 6.1–6.24), centric sSMC of unknown origin have also been reported. First, there are those found and reported before the advent of molecular cytogenetics; in Liehr (2011a) there are over 100 such cases listed as acro-C-1 to acro-C-119 and u-C-1 to u-C-28. Additionally, 980 such sSMC were summarized by Liehr and Weise (2007). Also an sSMC carrier may acquire just the karyotype 47,XN,+mar. In these cases it is essential to find out more about the origin and genetic content of the sSMC to provide appropriate genetic counseling.

Additionally, some rare instances have been reported in which the chromosomal origin of the sSMC could not be clarified, even after the application of sophisticated molecular (cytogenetic) approaches. There are five sSMC derived from one of the acrocentric chromosomes; for additional cases, see Sect. 7.25. The acrocentric origin is known as these sSMC carry two short arms containing NORs. However, none of the centromeric probes specific for chromosomes 13/21, 14/22, 15, or 22 were able to induce a signal on the corresponding sSMC. So they can just be described as inv dup (acro)(p10). At the same time, these sSMC did not lead to clinical problems (Liehr 2011a).

Chapter 7
Neocentric Small Supernumerary Marker Chromosomes by Chromosome

Among sSMC reported in patients, neocentric ones constitute one of the smallest groups (Liehr et al. 2007a). The chromosomal distribution compared with that of centric sSMC is given in Fig. 1.4 (see Chap. 1). Neocentric sSMC have a centromeric constriction but no detectable alpha-satellite DNA. Thus, they are also called analphoid markers, which "carry newly derived centromeres (or 'neocentromeres') that are apparently formed within interstitial chromosomal sites that have not previously been known to express centromere function" (Choo 1997). Still, it is unclear how a neocentromere is really acquired and formed on an acentric fragment. Especially puzzling about neocentromeres is that they may also form on an otherwise normal chromosome (Liehr et al. 2010b).

An interesting issue regarding neocentric sSMC is that their presence can lead to no imbalance at all (e.g., case 13-N-p21.31/1-1; Liehr 2011a), or to a gain of one copy (e.g., case 11-N-qt22/1-1), two copies (e.g., case 08-N-pt23.3/1-1) or four copies (e.g., case 13-N-qt32/1-3) of the corresponding region present on the sSMC; even loss of one copy (e.g., case 13-N-q31.1/1-1; Liehr 2011a) is possible. The latter is mostly correlated with the so-called McClintock mechanism (see Sect. 9.2). Furthermore, neocentric sSMC can appear in all shapes mentioned in Chaps. 1 and 3. They can be minute-, ring-, or inverted-duplication-shaped.

Almost 100 neocentric sSMC have been reported and there are three major clinical groups among them. The smallest group is that with no clinical consequences (~5%), followed by a similarly small group with moderate to severe clinical consequences (~7%), and then the largest group with the most severe clinical consequences. If a neocentric sSMC does not cause an imbalance (see above) and/or the sSMC causing a large imbalance is present only in a mosaic, clinical signs may be absent or mild.

If a neocentric sSMC carrier has clinical signs, these can be due to the gain of the sSMC spanning region, but they can also be due to loss of the sSMC from a previously balanced karyotype. Another aspect which has been hardly studied up to now in neocentric sSMC is the UPD problem of the sSMC's sister chromosomes. Whether such a UPD is present in these cases and whether it might be causative for (some of the) symptoms is not yet known.

T. Liehr, *Small Supernumerary Marker Chromosomes (sSMC)*,
DOI 10.1007/978-3-642-20766-2_7, © Springer-Verlag Berlin Heidelberg 2012

In summary, in more than 90% of cases the diagnosis of a neocentric sSMC is correlated with an adverse clinical outcome. This is mainly due to the sheer size of the imbalance, and is largely independent of the chromosomal region from which the sSMC is derived. Nonetheless, in ~10% of cases no or mild symptoms can be present in a carrier of a neocentric sSMC. If this is valid, in most cases this is caused either by a mosaic or by a balanced cytogenetic situation. These facts have to be considered by clinicians involved in genetic counseling of prenatal patients with de novo neocentric sSMC. During the lifetime neocentric sSMC sometimes tend to disappear, especially in peripheral blood cells. Thus, skin fibroblasts may be studied if a suggested low-level mosaicism with sSMC of unknown origin is suspected (Liehr 2011a).

7.1   Chromosome 1: Neocentrics

In clinical cases six neocentric sSMC(1) have been reported. Two were derived from the short arm and four from the long arm of chromosome 1, and were as shown in Fig. 7.1. Each neocentric sSMC(1) was a unique incident.

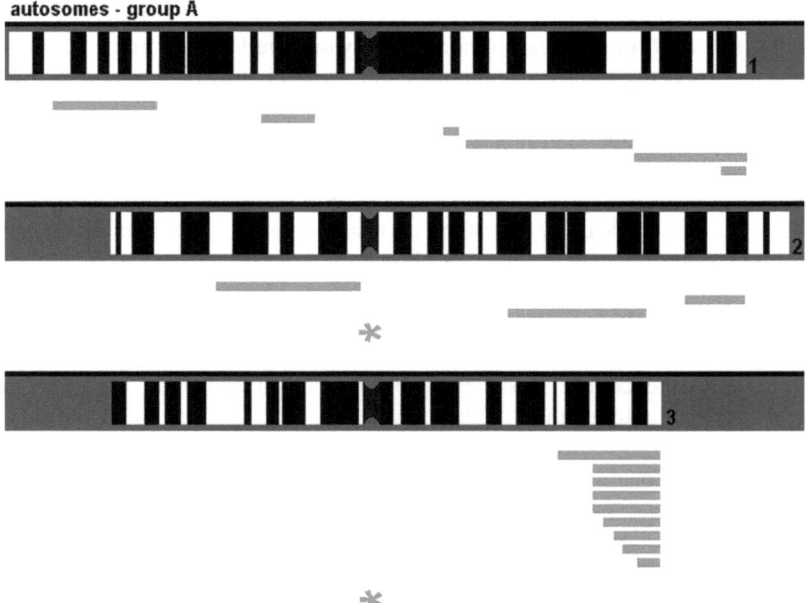

Fig. 7.1 Ideograms of chromosomes 1–3, i.e., the A group of the human karyotype; short arm on the *left*; long arm on the *right*. Below each ideogram *gray bars* are shown. Each bar stands for one neocentric derivative chromosome and *bar lengths* mark the region covered by the corresponding sSMC. *Asterisks* stand for a reported neocentric chromosome without known size or location of the sSMC

Interestingly, in one additional case (01-N-qt23~24/1-1; Liehr 2011a) the patient acquired an inv dup(1)(qter->q23~24::q23~24->qter) only in the malignant bone marrow cells (case not included in Fig. 7.1). N.B.: Five more exclusively tumor associated neocentric sSMC have been reported (see Sects. 7.3 and 7.9). According to present knowledge, this finding does *not* mean that (neocentric) sSMC carriers are more prone to any kind of cancer than the average population.

7.1.1   Balanced Situation

Two of the neocentric sSMC(1) led primarily to a balanced situation in the carriers (McClintock-mechanism; see Sect. 9.2). One of them (case 01-N-p32/1-1; Liehr 2011a) was reported to be clinically normal, even though secondarily in two further subpopulations the sSMC was duplicated or lost. The second was a prenatal case (01-N-q23/1-1; Liehr 2011a) and the parents opted for termination of the pregnancy, as variants of sSMC were detected, which led most probably to an unbalanced situation in some of the cells.

7.1.2   Unbalanced Situation

Even though descriptions of the remaining four neocentric sSMC(1) cases are rather scrappy, partial deletions in connection with sSMC seem not to be present. In fact, partial gain of copy number was observable and led to different clinical abnormalities.

7.2   Chromosome 2: Neocentrics

Four neocentric sSMC(2) have been reported, two derived from the long arm, one from the short arm, and one of unknown origin and size. No regions were overlapping or observed more frequently than once (Fig. 7.1).

7.2.1   Balanced Situation

Two of the four neocentric sSMC(2) were primarily formed by the McClintock mechanism (see Sect. 9.2) and led to a balanced situation. As both cases were connected with mental retardation, either the sSMC was a coincidental finding or, maybe more likely, sSMC(2) was lost or duplicated in tissues other than the ones studied and this led to symptom-causing imbalances there (cases McCl-02-N-p21/1-1, McCl-02-N-q35/1-1, Liehr 2011a).

7.2.2   Unbalanced Situation

The other two neocentric sSMC(2) were (low) mosaic in the tissues studied, and the sSMC(2) were there in addition to an otherwise normal karyotype. In case 02-N-1 (Liehr 2011a) 12% of amnion cells had sSMC(2) and there was a normal clinical outcome. However, in case 02-N-q22/1-1 (Liehr 2011a) sSMC(2) was present in 28% of the peripheral blood cells and the patient was dysmorphic with psychomotor retardation. Thus, most likely in the latter case the sSMC was present at a higher percentage in other body tissues of the patient.

7.3   Chromosome 3: Neocentrics

Ten neocentric sSMC(3), nine of which were derived from the distal tip of the long arm of chromosome 3, have been reported. In one additional case, the exact origin and size of the sSMC(3) was not reported. Not depicted in Fig. 7.1 are four additional neocentric sSMC derived from chromosome 3 observed exclusively in either leukemia, including Fanconi anemia (cases 03-N-qt26/2-1 to 03-N-qt26/2–2 and 03-N-qt26.3/1-1; Liehr 2011a), or a lung sarcoid carcinoma (case 03-N-qt26/1-1; Liehr 2011a). The latter case also had a tumor-associated neocentric sSMC(9) – see Sect. 7.9. N.B.: According to present knowledge, this finding does *not* mean that (neocentric) sSMC(3) carriers are more prone to any kind of cancer than the average population.

7.3.1   Balanced Situation

No balanced situation has been reported in neocentric sSMC(3).

7.3.2   Unbalanced Situation

All molecular-cytogenetically characterized neocentric sSMC(3) are reported to be inverted duplication marker chromosomes. They induce different sizes of terminal partial tetrasomy of the distal part of the long arm of chromosome 3. This leads to a relatively typical cluster of symptoms as listed in Table 7.1. Most typical are, besides dysmorphic features and mental retardation, a streaky hyperpigmentation following the line of Blaschko, kidney problems, and polydactyly. Similar signs were seen in a case with triplication of 3q25.3 to 3q29 (case 03-N-IMB-q25.3/1-1; Liehr 2011a).

Table 7.1 Symptoms regularly observed in neocentric ssMC(3) – distal long arm

Symptoms	3qter (%)
Brain malformations	27
Clubfeet	18
Dysmorphic face	64
Growth retardation	18
Heart defect	18
Kidney problems	36
Mental retardation	45
Macrocephaly	27
Polydactyly	27
Seizures	18
Streaky hyperpigmentation	55

According to 11 informative cases from Liehr (2011a)

7.3.3   Case Report

Provided by Unique; reported by the grandparents of a 12-year-old girl with karyotype 47,XX,+inv dup(3)(qter->q26.1::q26.1->qter)[4]/46,XX[26].

Twelve years ago we would never have been able to envisage the long road that our granddaughter Lauren [Note: most of the children's names have been changed in accordance with their parents' wishes] has traveled. She has already had over 50 surgical procedures in her short life... (Fig. 7.2)

Fig. 7.2 Lauren, who has a neocentric ssMC derived from chromosome 3 (copyright Unique)

They say that we have a sixth sense. The day Lauren was born we just knew that we had to be there outside the delivery suite. We had to be there to tell her that all would be well and that the love of her families would pull her through.

Lauren was born with a rare genetic defect involving a partial duplication of chromosome 3q which was diagnosed when she was 2 years and 7 months old. Although it was comforting in a strange way to have a diagnosis, the diagnosis

meant absolutely nothing to us or indeed to most of the medical consultants who were treating her. The geneticist told us that very little was known and he referred us to Unique, the Rare Chromosome Disorder Support Group based in England. Unique was and is our lifeline.

Parents of children who are born with more common genetic variations such as Down's syndrome are given a lot of information about the syndrome and usually have some idea about how their child will be affected medically and emotionally. They have support groups for both children and parents. But a child like Lauren – and that's the problem, there is no other child like her – has to find her own way in life and is really a medical first where each medical anomaly is dealt with as a separate entity.

At 2 years of age and after suffering 2 years of not being able to pass a bowel movement, her health was rapidly declining, she was losing weight and her hair was falling out. Lauren had an operation to give her a stoma so that we could insert a catheter into the stoma and fill her bowel with water to wash out the feces, an antegrade continence/colonic enema. She was the youngest patient in Ireland to have this procedure and we believe that it saved her life. Today, at the age of 12, Lauren is in nappies as she has no bladder control.

Once Lauren recovered from that operation she started to thrive and the doctors looked at her other problems. Today Lauren wears very strong glasses – R +5.00DS; L6.500 D and has bilateral strabismus. The ophthalmologist says that surgery will not make much difference. She has no lacrimal apparatus which used to cause a lot of eye infections but they have not been too bad recently.

She has constant ear infections and abscesses and both ear drums were perforated although the right ear has healed and she is now wearing a hearing aid in this ear.

Lauren attends a mainstream school with a full time special needs attendant and is approximately 2 years behind her peers. She wants to learn, is much better at spelling and English than at math, has problems with eye scanning but does have a very good memory. Her handwriting is not well formed and she finds it hard to keep the letters uniform. She loves music, especially Billy Joel.

We take good care of Lauren's teeth. She will probably need a brace on them in the future: she is always smiling and she has the most beautiful smile!

Lauren took her first tentative steps at about 2 years and 7 months. She was born with a left talipes which was splinted or plastered from birth. She twice had corrective surgery to lengthen her left Achilles tendon. Her feet and toes are odd sizes and shapes and she has had both legs put in plaster to try to straighten them. Lauren used the fact that her legs were in plaster to help her to stand up and walk. She has no muscles from the knees down and therefore prior to getting the plasters on she didn't have the strength to pull herself up. The plasters gave her lower legs an unexpected rigidity, and you can imagine the excitement when she took her first "steps" wearing her orthotic boots. Our little girl was delighted with herself.

Although she has the use of a wheelchair, Lauren can walk short distances on very flat surfaces and can negotiate stairs. She is much steadier going upstairs, usually one at a time, than coming downstairs. You have to keep reminding her to pay attention and to hold onto the banister tightly, a more frightening experience for us than for her!

We have only mentioned some of Lauren's many problems. Our philosophy is to take one day at a time with her and to give her as much love as is humanly possible. She is such a warm, loving little girl who does her best to make us all laugh.

Of course things are never "normal" with Lauren. When she was 5 years old we spotted a small red growth at the entrance of her vagina. At first the doctors were not worried but eventually after much prompting they investigated it. A considerable-sized müllerian papilloma was diagnosed which had to be excised three times and is still under scrutiny. Lauren's surgeon says that the growth is like palm fronds but that it is benign.

We have always been worried about Lauren's kidneys. From birth she suffered constant urinary tract infections and had protein in her urine. When she was 8, we were told that she had cysts on her kidneys but that they were just being monitored.

Friday 11th July is a day that we will never ever forget. That's the day, just 10 days before Lauren's eleventh birthday, that we were told that she had chronic renal failure and would need to go on dialysis awaiting a kidney transplant. After all that our little girl had endured, what more would she have to endure?

Of course as usual, Lauren took it all in her stride. She had the operation to give her a central line for dialysis, which she went to three times a week. Lauren found dialysis hard: through it all she would suffer from severe cramps and vomiting, but the only thing that she complained about was the loss of her planned holiday to Portugal.

Then joy of joys: after an 18-month wait, Lauren – now 12 years, 5 months old – got the call that we had all been waiting and hoping for. But instead of cries of happiness, we were shocked, frightened and elated – all at the same time.

The kidney took 7 days to kick in after transplant and Lauren suffered a bad stomach bleed because of ulcers. Two months on, the kidney is functioning but her magnesium levels are not stable. The doctors said that because she has a new kidney, we may now be able to toilet train her so that she can discard her nappies. We thank God for donors.

Lauren has lots of friends. She is a great mixer at school and she loves people, but she does get very frustrated because there are things that she just can't do. She loves water but can't go swimming because of her ears – so she compensates by asking us to fill the bath up "like a swimming pool".

She has very little social life outside of school, as there are no clubs who cater for her needs, but she never complains.

She is such a joy, she has suffered so very much in her short life, yet she only ever shows a happy exterior. People who meet Lauren never forget her, her wit and her non-stop talking. She is a shining example to us all.

That's our Lauren – predictably unpredictable! We love her.

7.4   Chromosome 4: Neocentrics

There is only one case report available on a neocentric sSMC(4) – see Fig. 7.3.

In this case the sSMC was formed by the McClintock mechanism (see Sect. 9.2). The originally balanced karyotype lost sSMC(4), and thus led to partial monosomy

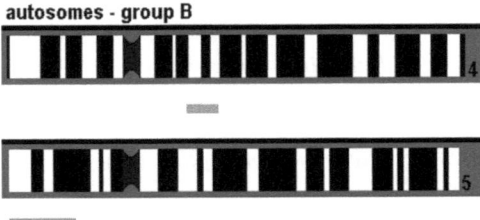

Fig. 7.3 Ideograms of chromosomes 4 and 5, i.e., the B group of the human karyotype; short arm on the *left*; long arm on the *right*. Below each ideogram *gray bars* are shown. Each bar stands for one neocentric derivative chromosome and *bar lengths* mark the region covered by the corresponding sSMC

in some of the carrier's cells. It would be impossible to decide if the clinical symptoms in this patient were connected with sSMC formation as the patient's mother (without karyotypic abnormalities) was diagnosed with borderline mental retardation and autism.

7.5   Chromosome 5: Neocentrics

As for chromosome 4, also only one carrier of a neocentric sSMC(5) is known (Fig. 7.3). However, this patient had an sSMC(5) leading to partial tetrasomy 5pter to 5p14 and showed clinical symptoms very similar to those that patients with an isochromosome 5p show (see Sect. 6.5.2.1). Thus, the "isochromosome 5p syndrome" critical region can be narrowed down to 5pter to 5p14 instead of the whole short arm of chromosome 5.

7.6   Chromosome 6: Neocentrics

For chromosome 6 only two neocentric sSMC are known – both derived from the long arm (Fig. 7.4).

7.6.1   Balanced Situation

Case 06-N-q16.2/1-1 (Liehr 2011a) involved a patient with neocentric sSMC(6) but there were no clinical signs, as the marker presence was due to a McClintock mechanism (see Sect. 9.2) based balanced rearrangement in all cells studied. However, this carrier was the father of an affected child with the same balanced rearrangement in all cells studied.

7.6.2   Unbalanced Situation

For case 06-N-qt26/1-1 (Liehr 2011a) a terminal partial tetrasomy was induced by the neocentric sSMC(6). This imbalance was noted prenatally and was primarily present in amnion and placenta, but was not detectable in peripheral blood. The patient was reported as clinically normal at 2 years with mild hearing loss.

7.7   Chromosome 7: Neocentrics

Even though two neocentric sSMC(7) have been reported, the details provided on these cases are sparse (Fig. 7.4). There are hardly any clinical details and/or details on the marker size and shape available. One patient had learning and developmental delay as well as dysmorphic features, but nothing is known about the sSMC(7) itself, apart from the fact that it was present in mosaic (case 07-N-mar/1, Liehr 2011a). The second case was a inverted-duplication-shaped sSMC(7) derived from the distal end of the chromosome's long arm, but no clinical data were provided (07-N-qt36.1/1-1; Liehr 2011a).

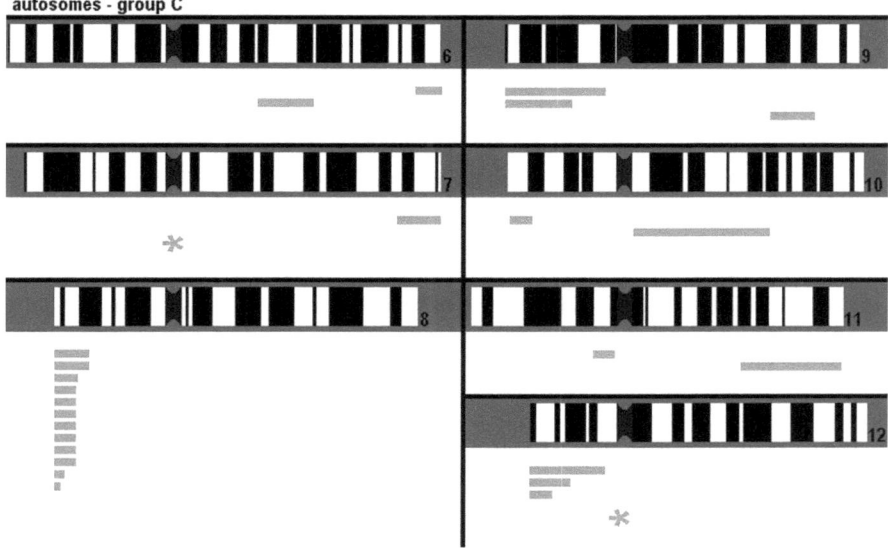

Fig. 7.4 Ideograms of chromosomes 6–12, i.e., the C group of the human karyotype; short arm on the *left*; long arm on the *right*. Below each ideogram *gray bars* are shown. Each bar stands for one neocentric derivative chromosome and bar lengths mark the region covered by the corresponding sSMC. *Asterisks* stand for a reported neocentric chromosome without known size or location of the sSMC

7.8   Chromosome 8: Neocentrics

As shown in Fig. 7.4 for chromosome 8, 12 neocentric sSMC are known. All of them led to unbalanced situations, i.e., partial tetrasomy of parts of the distal short arm. It seems that mosaicism, i.e., the size of the cell clone without sSMC(8), influences the clinical outcome; patients with a small percentage of cells with sSMC in the tissues studied tended to be less affected, e.g., cases 08-N-pt23.2~23.1/1-1, 08-N-pt23.1/1-4, and 08-N-pt22/1-1 (Liehr 2011a). However, in general, no unaffected carrier of a neocentric sSMC(8) is among the reported cases. Some of the carriers might resemble patients with isochromosome 8p (see Sect. 6.8.2.1) however, the critical region for this kind of "syndrome" may not be in the distal region of the short arm of the chromosome.

7.9   Chromosome 9: Neocentrics

Three constitutional neocentric sSMC(9) are known from clinical cases and one was acquired (not shown in Fig. 7.4) from a lung sarcanoid carcinoma (case 09-N-qt26/1-1; Liehr 2011a), the latter also having a tumor-associated neocentric sSMC(3) – see Sect. 7.3. Two of the constitutional sSMC(9) were from the short arm and the remainder were from the long arm of chromosome 9.

7.9.1   Balanced Situation

There is a case of one female without clinical signs (09-N-q31/1-1; Liehr 2011a) who had a neocentric sSMC(9) formed by the McClintock mechanism (see Sect. 9.2). Even though she had partial deletion of 9q31 to 9q33 in 40% of her peripheral blood due to loss of sSMC(9) in these cells, she was clinically normal. She was picked up as she gave birth to a son with the same deletion in all his cells, who was certainly mentally impaired because of that.

7.9.2   Unbalanced Situation

There are two cases leading to unbalanced situations; in both cases there is an inverted duplicated neocentric sSMC(9). In one case (09-N-pt21.1/1-1; Liehr 2011a) partial tetrasomy 9pter to 9p21.1 was the consequence, present in all cells studied. A phenotype similar to that in "isochromosome 9p syndrome" was observable (see Sect. 6.9.2.1), narrowing down the critical region for this syndrome to 9pter to 9p21.1.

In case 09-N-pt12/1-1 (Liehr 2011a) there was also an inverted duplicated chromosome including 9pter to 9p12. However, in this case there was also a deletion on one of the two centric chromosomes 9, i.e., the identical region 9pter to 9p12 was absent there. Thus, the inverted duplicated neocentric sSMC(9) on the one hand compensated for the loss and additionally provided partial trisomy of this region. Symptoms similar to those in trisomy 9p syndrome, in general, were the consequence in this case (see Sect. 6.9.2.1).

7.10 Chromosome 10: Neocentrics

Only two cases of neocentric sSMC(10) are available in the medical literature. In one case the neocentric sSMC(10) was derived from the short arm and in the other case it was derived from the long arm (Fig. 7.4).

7.10.1 Balanced Situation

An originally balanced situation caused most likely by the McClintock mechanism (see Sect. 9.2) was observed in a child with a neocentric sSMC(10). Nonetheless, the child showed mental retardation, most likely as the balanced karyotype was only present in ~60% of the cells studied; i.e., the sSMC was lost and this led to partial monosomy 10q11 to 10q23.

7.10.2 Unbalanced Situation

Another neocentric sSMC(10) patient had a distal partial terminal short arm tetrasomy. Even though the sonographic findings were normal, the pregnancy was electively terminated.

7.11 Chromosome 11: Neocentrics

As for chromosome 10, only two neocentric sSMC(11) are known, one derived from the short arm and the other derived from the long arm (Fig. 7.4).

7.11.1 Balanced Situation

One neocentric sSMC(11) was due to the McClintock mechanism (see Sect. 9.2) and was found in a clinically normal female in all cells studied (case 11-N-p11.2/1-1; Liehr 2011a). It was detected because of a clinically abnormal child who had an unbalanced karyotype because of the maternal chromosomal rearrangement.

7.11.2 Unbalanced Situation

As in case 09-N-pt12/1-1 (Liehr 2011a), the reported neocentric sSMC(11) was an inverted-duplication-shaped derivative which partially compensates for a deletion present in one of the corresponding centric sister chromosomes. Thus, in case 11-N-qt22/1-1 (Liehr 2011a) the sSMC led to partial trisomy 11q22 to 11qter and expected clinical signs such as dysmorphism and mental and developmental delay.

7.12   Chromosome 12: Neocentrics

As shown in Fig. 7.4, four neocentric sSMC derived from chromosome 12 are known. For one of them, the size and the region covered are unknown (case 12-N-mar/1; Liehr 2011a); the other three all derive from the distal part of the short arm of chromosome 12. All of them, including case 12-N-mar/1 (Liehr 2011a), led to clinical signs comparable to those observed in PKS. Thus, all four of them derived most likely from the short arm of chromosome 12 and were inverted-duplication-shaped. According to the smallest of these neocentric sSMC(12) (case 12-N-pt11.22/1-1; Liehr 2011a), the PKS-critical region must be somewhere between 12pter and 12p11.22. Interestingly, as is also typical for PKS, in two of these neocentric sSMC cases the sSMC was absent from peripheral blood.

7.13   Chromosome 13: Neocentrics

There are 15 reports of neocentric sSMC for chromosome 13. Two of them are ring-shaped, and the remaining ones are inverted-duplication-shaped. As chromosome 13 is an acrocentric chromosome, all neocentric sSMC are derived from the long chromosome arm (Fig. 7.5).

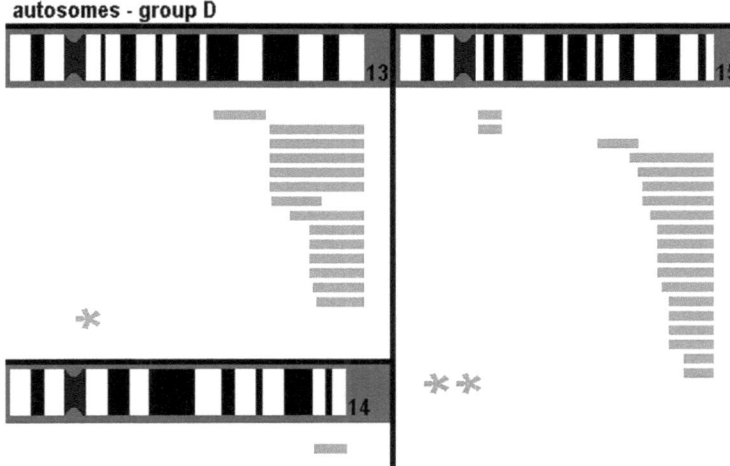

Fig. 7.5 Ideograms of chromosomes 13–15, i.e., the D group of the human karyotype; short arm on the *left*; long arm on the *right*. Below each ideogram *gray bars* are shown. Each bar stands for one neocentric derivative chromosome and bar lengths mark the region covered by the corresponding sSMC. *Asterisks* stand for a reported neocentric chromosome without known size or location of the sSMC

7.13.1   Balanced Situation

Two reports on neocentric sSMC(13) with an originally balanced karyotype due to marker chromosome formation by the McClintock mechanism (see Sect. 9.2) are available. In case 13-N-p21.31/1-1 (Liehr 2011a), the sSMC was in all of the cells of a normal female that was studied because of repeated abortions. Case 13-N-q31.1/1-1 (Liehr 2011a) also involved an originally balanced karyotype, but the sSMC was lost in a considerable number of the cells, leading to partial monosomy and in consequence to clinical signs and symptoms.

7.13.2   Unbalanced Situation

All remaining 13 neocentric sSMC(13) were inverted-duplication-shaped extra chromosomes. They led to partial trisomy (case 13-N-qt32.3/2-1; Liehr 2011a), tetrasomy (most cases), and even octasomy (case 13-N-qt32.3/1-1; Liehr 2011a). As seen in Table 7.2, neocentric sSMC(13) derived from the distal long arm led to seizures, brain malformation, and overgrowth as well as dysmorphism and mental retardation.

Table 7.2 Symptoms regularly observed in neocentric ssMC(13) – distal long arm

Symptoms	13qter (%)
Brain malformations	33
Dysmorphic face	100
Heart defect	12
Mental retardation	67
Microcephaly	12
Overgrowth	22
Scoliosis	12
Seizures	56
Strabismus	22

According to nine informative cases from Liehr (2011a)

7.14   Chromosome 14: Neocentrics

One neocentric inverted-duplication-shaped ssMC(14) inducing partial trisomy 14 is known (Fig. 7.5). As described for a neocentric ssMC(9) before (see Sect. 7.9.2) the inverted duplicated chromosome is associated with a partial deletion of the same region on one centric sister chromosome 14. As present in all cells, the imbalance led to appreciable clinical problems in the carrier (Liehr 2011a).

7.15   Chromosome 15: Neocentrics

Among the neocentric ssMC, those derived from chromosome 15 are the most frequently observed. They constitute ~25% of this group. Remarkably, this is a proportion similar to that in the group of centric ssMC. Most of the neocentric ssMC(15) are inverted-duplication-shaped derivative chromosomes. As outlined in Sect. 3.1, Murmann et al. (2009) recently provided evidence for a different mechanism in the formation of neocentric and centric inverted-duplication-shaped ssMC. But it is still a strange coincidence that in these two groups specifically chromosome 15 is most frequently involved in ssMC formation.

Overall, 22 neocentric ssMC(15) have been reported. Two were not characterized in detail for the region of origin, three were ring-shaped, and the remainder were inverted-duplication-shaped. All were derived from the long chromosome arm, mainly from the distal end (Fig. 7.5) (Liehr 2011a).

7.15.1   Balanced Situation

Even though neocentric ssMC(15) are seen most often, there is no case of a patient with a balanced situation, which would be due to the McClintock mechanism (see Sect. 9.2).

7.15.2   Unbalanced Situation

7.15.2.1   Ring-Shaped Neocentric sSMC

Two of the three ring-shaped neocentric sSMC(15) are very much alike: cases 15-N-q11.2/1-1 and 15-N-q11.2/1–2 (Liehr 2011a) led to partial tetrasomy of 15q11.2 to 15q13. Both carriers had relatively mild symptoms, including delayed puberty.

The third ring-shaped neocentric sSMC(15) led to partial trisomy of 15q22.1~22.2 to 15q22.3~24.1 and there is a single case report with dysmorphism, prenatal and perinatal growth retardation, altered to overgrowth and obesity around puberty similarly as in PWS; however, the male patient was reported to be mentally normally developed in the observation period between 15 and 21 years of age (case 15-N-q22.1/1-1; Liehr 2011a).

7.15.2.2   Inverted-Duplication-Shaped Neocentric sSMC

The remaining well-characterized neocentric sSMC(15) were inverted-duplication-shaped and led to partial terminal tetrasomy 15q. Clinically typical are dysmorphism, mental impairment, overgrowth, hip problems, sensorineural hearing loss, kidney malformations, and heart problems (Table 7.3). The critical region seems to be in 15q25~q26.

Table 7.3 Symptoms regularly observed in neocentric sSMC(15) – distal long arm – with clinical symptoms

Symptoms	15qter (%)
Dysmorphic face	86
Hearing loss (sensor neural)	29
Heart defect	21
Hip problems	29
Hypotonia	14
Kidney malformations	29
Mental retardation	79
Overgrowth	57

According to 14 informative cases from Liehr (2011a)

7.16   Chromosome 16: Neocentrics

For chromosome 16, two neocentric sSMC have been reported (Fig. 7.6). One was not characterized in detail (case 16-N-mar/1; Liehr 2011a). Case 16-N-p11.2/1-1 (Liehr 2011a) had overall complete trisomy of chromosome 16, which is known to be lethal prenatally. Here a centric isochromosome 16q and an acentric isochromosome 16p were present together with a normal chromosome 16.

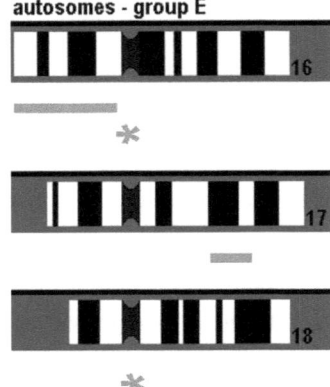

Fig. 7.6 Ideograms of chromosomes 16–18, i.e., the E group of the human karyotype; short arm on the *left*; long arm on the *right*. Below each ideogram *gray bars* are shown. Each bar stands for one neocentric derivative chromosome and *bar lengths* mark the region covered by the corresponding sSMC. *Asterisks* stand for a reported neocentric chromosome without known size or location of the sSMC

7.17 Chromosome 17: Neocentrics

The only known neocentric sSMC(17) had an unusual shape and mode of formation compared with other neocentric sSMC: an inverted-duplication-shaped sSMC was present which was not derived from either distal end of the chromosome. In contrast, it evolved from an interstitial deletion of the region 17q22 to 17q23 of one of the two centric chromosomes 17 (Fig. 7.6). Thus, overall this sSMC led to a partial trisomy, even though it was present with an inverted-duplication-shaped sSMC. As to be expected, multiple congenital abnormalities were connected with this aberrant karyotype.

7.18 Chromosome 18: Neocentrics

Only one, not well-characterized neocentric sSMC(18) is known (Fig. 7.6). According to the clinical symptoms – see case 18-N-mar/1 (Liehr 2011a) – one can speculate whether the sSMC might be derived from the short arm of chromosome 18.

7.19 Chromosome 19: Neocentrics

For chromosome 19 neither a neocentric sSMC nor a neocentromere formation with another rearrangement has ever been seen (Fig. 7.7).

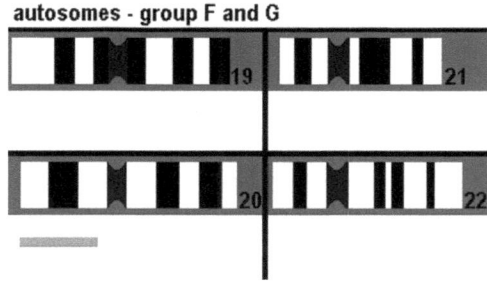

Fig. 7.7 Ideograms of chromosomes 19–22, i.e., the F group and the G group of the human karyotype; short arm on the *left*; long arm on the *right*. Below each ideogram *gray bars* are shown. Each bar stands for one neocentric derivative chromosome and *bar lengths* mark the region covered by the corresponding sSMC

7.20   Chromosome 20: Neocentrics

One neocentric sSMC(20) derived from the short chromosome arm is known (Fig. 7.7). The inverted-duplication-shaped sSMC leads to a partial trisomy in the same way as outlined in Sect. 7.9 for a neocentric sSMC(9). The case shows the typical partial trisomy 20p phenotype (Schinzel 2001; see Sect. 6.20.2.1).

7.21   Chromosome 21: Neocentrics

No neocentric sSMC(21) is known (Fig. 7.7); however, chromosome 21 can form a neocentromere at least at one locus as shown by Barbi et al. (2000).

7.22   Chromosome 22: Neocentrics

As reported for chromosome 19, in chromosome 22 no neocentromere formation has been observed (Fig. 7.7).

7.23   X Chromosome: Neocentrics

For the X chromosome, the first neocentric sSMC was found recently (case-0X-N-pt22.31/1-1; Liehr 2011a). The inverted-duplication-shaped sSMC led to partial tetrasomy Xp22.31 to Xpter (Fig. 7.8) and severe physical and mental impairment in the carrier.

gonosomes

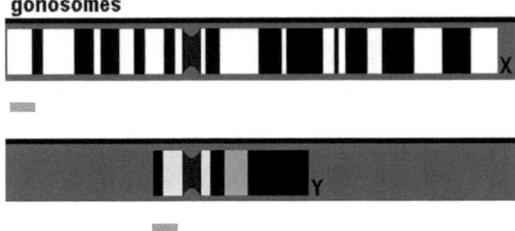

Fig. 7.8 Ideograms of the X and Y chromosomes, i.e., the gonosome group of the human karyotype; short arm on the *left*; long arm on the *right*. Below each ideogram *gray bars* are shown. Each bar stands for one neocentric derivative chromosome and bar lengths mark the region covered by the corresponding sSMC

7.24 Y Chromosome: Neocentrics

Even though more than five cases with Y chromosome neocentromere formation are known, only one of them was an sSMC (Fig. 7.8). It was detected prenatally and the parents elected for termination of the pregnancy.

7.25 Neocentric sSMC of Unknown Origin

As mentioned in Sect. 6.25, there are six sSMC derived from one of the acrocentric chromosomes described as inv dup (acro)(p10). Recently we came across a cytogenetically similar case of an sSMC carrying two short arms containing NOR and stainable with none of the centromeric probes specific for chromosomes 13/21, 14/22, 15, or 22. Here additionally we tested for centromeric activity and found the sSMC to behave like a neocentric chromosome (case acro-N-mar/1; Liehr 2011a). Thus, either some or all inv dup(acro) sSMC are neocentric sSMC.

Second, there are two sSMC cases which are even weirder. In these cases, the sSMC were not stainable by any available human DNA probe. They could only be stained by DNA derived from the sSMC itself (Liehr et al. 2008a; see Sect. 2.2) and must be neocentric. It can be speculated that any repetitive sequences were amplified, and formed an sSMC for unknown reasons. As in one of the two cases no information and in the other case only sparse clinical information is available, no genotype–phenotype correlation can be made.

Chapter 8
Multiple Small Supernumerary Marker Chromosomes

If an sSMC carrier has not only one (or more identical) centric sSMC derived from only one chromosome [e.g., case 09-O-p12/6-1 with up to five sSMC(9)], but has two or more sSMC, each derived from different chromosomes, one speaks of "multiple sSMC." The chromosomal distribution of multiple sSMC is given in Fig. 1.3 (see Chap. 1). Most of the carriers of multiple sSMC are clinically abnormal and often severely impaired; there are only a few exceptions showing minor or no clinical signs or symptoms (mult 2-14, mult 2-17, mult 2-23, mult 2-29, mult 3-3, mult 6-1, mult 7-1, mult 7-2; Liehr 2011a).

Overall, multiple sSMC can be harmless if they cover only potentially non-dose-sensitive pericentric regions as defined in Chap. 6. The clinical signs seem to be like those reported in Chap. 6 for the corresponding chromosomal regions.

Mosaicism is frequent in multiple sSMC, even providing cell lines with different marker chromosomes besides those where all sSMC are present in one cell together. Studies of UPD are sparse in multiple sSMC. Thus, the influence of this feature in the symptoms is unknown. For formation of multiple sSMC, see Sect. 3.6.

8.1 Case Reports

Report A: *Provided by Unique; reported by the mother of a 14-year-old boy with karyotype 48,XY,+r(X),+r(17)/47,XY,+r(X)/47,XY,+r(17).*

Tom [Note: most of the children's names have been changed in accordance with their parents' wishes] is our eldest child of four, and was born in 1997, at term after a normal pregnancy. It was a relatively fast labor (5 hours) which ended in a ventouse delivery because the cord got wrapped around his neck and his heart rate dropped. We took him home after a few days, thinking all was fine, apart from having a toe on his right foot which crossed another one. Looking at photos now it is obvious that he was puffy and did not look like a pink healthy baby, but hindsight is a great thing. I took him for a walk one day and when we got home his feet were blue, so I went and bought him some thermal leggings. However the problem was not cold, it was the fact that he was in heart failure and when he was 12 days old he crashed. He woke up blue that morning, so we took him into accident and emergency and within an hour he was intubated. That evening we got flown up to Greenlane hospital in Auckland, the main center in New Zealand for pediatric heart surgery at that time, and he had a coarctation repair the next day. We were told that he should make a quick recovery, but he had problems gaining weight and took 6 weeks to get back to his birth weight, so we were in hospital for about 4 weeks, and he never really mastered breastfeeding.

After this initial hiccup we thought all was well. I think if Tom had not been my first child I would have realized that he was floppier than he should have been, and that his difficulty with solid foods was not normal. We really started to realize that all was not right when every other baby in my antenatal group was sitting up and he was not. At about 10 months we started with a speech therapist for feeding since he would only eat sweet things that came in jars, physiotherapy and occupational therapy. At about this age he also became more difficult to put to bed at night, a nurse suggesting that this was perhaps because he was not as physically active as a baby who was doing all the things he should be doing. He slept through once he went to sleep, but he often was not asleep until 10.00 in the evening and needed someone to sit with him until he went to sleep – this process usually took about an hour and a half. He got slightly easier to put to bed when he went to school.

Tom sat at 10 months, crawled at 17 months and walked at 26. Eating continued to be a problem. Until he was 2½ he would chew finger food but then spit it out and would only swallow purées. His speech was very slow to arrive and his first words at about 2 were "yoghurt" and "beer" – not exactly normal speech patterns! He also had a minor form of hypospadias which necessitated a circumcision. All these things led the pediatrician to do a chromosome test when Tom was just over 2. It came back abnormal. It was good to know there was a reason for all the things that were wrong with our boy, but as his karyotype appears to be unique, it does not provide any answers. What was perhaps more of a shock was when they tested us, and my karyotype also came back abnormal – 3% of my cells have extra 17th chromosome material in them. The geneticists were not sure if there was a direct connection between my abnormality and Tom's but wondered if there was a predisposition for my genetic material to fracture.

Although Tom kept making progress, his development was well behind "normal" in many areas, and when he was 3 we were told that Tom was on the autistic spectrum. This is the aspect of Tom that has the largest impact on our daily lives.

Tom attends mainstream school, with full time teacher aide support, funded by the state. He is currently in Year 8 at our local full primary school, and will go on to secondary school (high school/college) when he is 14. He can be very disruptive in the classroom, yelling and screaming when he does not want to do something, but probably does this less often now than in the past. Mainstream school is the right place for him because he needs and enjoys interaction with normal children. Over his years at primary school his ability to be integrated into the classroom program has increased – when he started he could not even sit on the mat! He always has his head in a book, and although it is hard to test comprehension, he is currently

Fig. 8.1 Tom at the age of
12 years (copyright Unique)

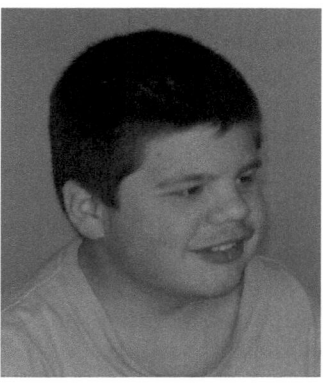

*reading and understanding at about 8 year old level. Tom can add and do simple
subtraction. He loves the computer, and as he has not got the physical co-ordination
to write neatly, the day he mastered the mouse was an important milestone indeed!
Tom can talk in simple sentences, mainly to ask for something he wants, but his
ability to make himself understood is improving all the time (Fig. 8.1).*

*Tom loves the television and has a general aversion to most forms of physical
exercise, probably due to his low muscle tone and lack of coordination. He enjoys
swimming in both the pool and ocean, and can walk long distances, especially if
incentivized by a trip to the bookshop for a comic, or time on the computer. He has
always been short, but is also rather round as well, despite our efforts to watch his
diet. We suspect that left to his own devices he would just keep eating, rather than
knowing to stop when he was full. However, at the age of 13 he is well advanced
through puberty and he seems to be maintaining a steady height/weight ratio.*

*He is still monitored by cardiologists, but has required no intervention since his
initial repair. He has no other health issues, and is on no medication. One major
milestone was when he became toilet trained, when he was about 9 during the day,
and a year later at night. That was something I had thought might never happen!*

*I have always been grateful that Tom copes with change and new situations
relatively well, unlike many autistic kids. Although he can be a bit of a handful out
in public, as he will do things like go up to people and want to read their T shirts, he
enjoys getting out and about. He loves going to bookshops or electrical shops and
looking at all the things and he is getting better at leaving when we say! He likes
going to the local tourist offices (information centers) and getting the free accom-
modation/tourist brochures, which he spends hours reading, and he really likes
planning trips to new towns.*

*The children he has been at school with are wonderful, and Tom is really well
liked and accepted at school, but he does not have friends as such. Secondary
school will be a big change, and we are still to make a decision as to where he will go.
For years it has been a prospect too scary to think about, but now I think he is
ready. The focus over the next few years will be on gaining life skills and increasing
his independence as much as possible. Although I am sure the future will bring*

many challenges, I have learned not to look ahead too much, not to stress about things before I need to, and to enjoy every small achievement!

Report B: Provided by Unique; reported by the mother of a 27-year-old young female with karyotype 50,XX,+4mar[5]/49,XX,+3mar[4]/48,XX,+2mar[11].ish +mar1(X)(wcpX+),+mar2(3)(wcp3+).+mar3(8)(wcp8+).+mar4(?).

It was obvious when Heather [Note: most of the children's names have been changed in accordance with their parents' wishes] was born 27 years ago that all was not entirely well and she spent her first 10 days in the neonatal intensive care unit, with 10 further days in the special care unit. She was on the small side weighing 5 pounds, 12 ounces (just 2.6 kg) but dropped to 5 pounds 4 ounces (2.38 kg) after birth. She had seizures while I was pregnant and soon after birth. While in the hospital, she was diagnosed with neuromuscular scoliosis, 11 sets of ribs, hypotonia and dislocated hips for which she wore a splint until 4 months of age. She had bilateral strabismus and a blocked left tear duct. She also looked dysmorphic with disproportionate microcephaly, a prominent nasal bridge, a broad nasal root and midface hypoplasia. She has a shallow presacral dimple. At 11 months old she was diagnosed as having a trabeculated bladder and horseshoe kidneys. She has had ongoing difficulties feeding; even as an adult, her high arched palate and being hypotonic causes a tongue thrust when eating or drinking which leaves her at high risk for choking. She was born with severe reflux but was not diagnosed until the age of 14 and is controlled with omeprazole. She was diagnosed with hypothyroidism. Growth was extremely slow and was never on the growth chart. At 4½ years she was the height and weight of a 3-year-old child and as an adult is just 4 foot 6 (1.37 m) tall. Her head grew even more slowly, reaching the 50th centile for a 10-month-old baby when Heather was 4½ years old. Heather has outgrown some of her problems; others have needed intervention. A very soft grade 1 heart murmur was heard at 11 months; but at 4½ her heart was declared normal. Her trabeculated bladder changed to normal with time and functions fine and she has no renal dysfunction or urinary tract infections. The strabismus was corrected surgically but by adulthood, one eye or the other eye drifts due to only using one eye at a time (Fig. 8.2).

Heather's left foot was bent upwards at the ankle but has normalized with physiotherapy and an abductor release was done in her left groin. Nonetheless, her feet are very small and somewhat deformed: the middle toes are shorter than the

Fig. 8.2 Heather at different ages: 4 months (**a**), 11 months (**b**), 9 years (**c**), and 26 years (**d**) (copyright Unique)

other toes and her toenails grow slowly and turn up, while her big toes turn inwards as if she has bunions (which she does not).

As a baby, Heather kept her thumbs tucked into the palms and to some extent still does as a young adult. She does have palmar creases in both hands. She used to hold her left arm up bent at the elbow, with her right arm drooping. She has a loose side and a tight side. As an adult, her arms stay flexed at the elbows and she can neither straighten them nor raise them above her head. Heather's epilepsy was controlled by phenobarbital until the age of 3 years until she had multiple ear infections and high fever, when the seizures returned. She had PE tubes placed in her ears three times due to so many ear infections but eventually outgrew them. Investigations revealed an abnormal EEG and she was put on anti-epileptic medication – phenytoin, phenobarbital and eventually valproic acid. Heather's neuromuscular scoliosis was treated with a back brace to prevent surgery from the age of 9–18 and she did not have to have surgery.

Her physical development was delayed: she was slow to sit up, never did crawl but started walking at 3. She was called a "floppy baby" due to hypotonia. Today, as a young adult, she remains uncoordinated, clumsy and falls easily. Her left side is weak and hypotonia has persisted, so when walking, especially on unfamiliar ground, she needs her hand held. She also needs help with eating, having her food cut up and to be fed foods that she cannot pick up with a fork or spoon. She drinks out of a sippy cup.

Heather said her first words at the age of 2, and as an adult speech vocabulary consists of about 40 words and an approximate Signing vocabulary of around 50 words. As an adult she is a multimodality communicator, using words, sounds and Signs. She still has weekly speech therapy using preschool educational computer programs or doing table work with, for example, small toy animals and flash cards. She is very sensitive to loud or sudden noises and is child-like with a child-sounding voice, so people find it hard to believe she is in her twenties.

Heather's learning disabilities are moderate to severe and as an adult, her mentality is at the level of a 2- to 5-year-old. She remains in diapers and is totally dependent on others for her care, unable to bathe, brush her teeth, dress or toilet.

But it is important to remember that although Heather has so many problems, she is very healthy. Her many strengths: she has a good, equable disposition, has great sense of humor, is always in a good mood and is very, very sociable.

Chapter 9
Small Supernumerary Marker Chromosomes Additionally to Other Chromosomal Rearrangements

An sSMC can appear as a single chromosomal aberration or together with other genetic changes present in the patient. In the following, such cases are treated in detail. However, one case fitting none of the following groups is mentioned here, i.e. an exceptional case being a chimera of two cell lines with karyotype chi46,XYpat[18]/47,XX,+inv dup(13)[8]. The reason for spontaneous abortion in this case (13-U-4; Liehr 2011a) remains elusive.

9.1 Numerical Aberrations

9.1.1 TS Karyotype

An sSMC can be present in a "basic TS karyotype," i.e., only one X chromosome is present instead of two gonosomes in the patient's cells. The sSMC is then typically derived either from an X chromosome or from a Y chromosome (see Sects. 1.3.2 and 5.5). One case in which monosomy X was combined with trisomy 21 plus sSMC(X) was even reported (case min-p11.1/1-1; Liehr 2011a).

Rarely, there are autosomal sSMC combined with a basic 45,X karyotype (cases 14-U-13, 20-U-2, A-n.a./1-1, and A-n.a./1–2; Liehr 2011a). In addition, an sSMC can be present in one cell line with two gonosomes (i.e., 47 chromosomes in total) and in a second normal cell line (with 46 chromosomes). An example is case 02-O-p11.1/1-2 (Liehr 2011a) with karyotype 45,X[5]/47,XX,+mar[43]/46,XX[2]. Cases similar to this are cases 18-Wi-88, 18-Wi-157, 22-O-q11.21/1-4, 0X-CW-5, and 0X-U-6 (Liehr 2011a).

Overall, almost 600 sSMC have been found in various combinations with a "TS karyotype."

T. Liehr, *Small Supernumerary Marker Chromosomes (sSMC)*,
DOI 10.1007/978-3-642-20766-2_9, © Springer-Verlag Berlin Heidelberg 2012

9.1.2 Gain of Gonosomes

Other combinations of an sSMC with numerical gonosomal aberrations re less frequent than sSMC with from monosomy X (see Sect. 9.1.1). Four cases in combination with a triple X syndrome karyotype (09-O-pter/1-3, 09-U-16, 14-O-q11.1/1-5, and 14-U-5), five cases together with a Klinefelter syndrome karyotype (XXY condition; cases 06-CW-3, 07-U-6; 09-U-5, 0X-CW-10, and 0X-U-3, Liehr 2011a), and one case with an XYY syndrome karyotype have been reported (case 14/22-U-15; Liehr 2011a).

9.1.3 Trisomy 21

sSMC in combination with trisomy 21, i.e., Down syndrome, is a rare condition. Overall, around 40 cases are known (Liehr 2011a). For more than 12 of them the sSMC origin has been determined. Two sSMC were derived from nonacrocentric chromosomes (chromosomes 4, 7, and 16), the remainder originated from acrocentric chromosomes, chromosomes 15 and 22 not being the most frequently involved ones. Although this finding could result from the low number of cases, it supports the suspicion that, in general, the formation is different in sSMC patients with trisomy 21 compared with centric sSMC. As mentioned in Sect. 9.1.1, there is one case with trisomy 21, sSMC, and monosomy X (minX-p11.1/1-1; Liehr 2011a). For the clinical outcome, there was no hint that the sSMC influenced the symptoms significantly; however, this could be due to Down syndrome being a very variable clinical condition.

9.1.4 Other Autosomal Gain

Two cases with trisomy of an autosome other than chromosome 21 are known. One case with trisomy 15 was reported prenatally and had additionally an inv dup(13) (q11) or inv dup (21)(q11.1), i.e., case 13/21-U34 (Liehr 2011a). In the second case (15-U-164; Liehr 2011a) inv dup(15) was accompanied by trisomy 18. In both instances the trisomy of an entire chromosome was more important for the adverse clinical outcome than the presence of the sSMC.

9.2 Structural Aberrations

9.2.1 Microdeletions

Microdeletions together with an sSMC have so far only been seen in sSMC derived from chromosome 15, which is in all probability due to ascertainment bias.

First, there is one AS case and there are two PWS cases with sSMC(15) with a microdeletion in the PWS/AS critical region of chromosome 15   (cases 15-A-q11~12/1-1, 15-P-q11.1/2-1, 15-P-q11.1/2-3; Liehr 2011a; see Sect. 6.15.4). Second, in case 15-U-133 together with an inv dup(15)(q12) there is a microdeletion in the Williams–Beuren syndrome (OMIM #194050) critical region on 7q11.2.

These cases serve to remind us that the finding of an sSMC does not explain the clinical problems of the patient in every case. Comprehensive sSMC characterization must accompany a critical evaluation of and comparison with known cases with a similar sSMC and the case under consideration. If comparable sSMC cases do not show a comparable phenotype, the clinician might decide to look for another reason for the patient's signs and symptoms.

9.2.2   McClintock Mechanism

Around 30 sSMC were formed on the basis of the so-called McClintock mechanism. The way in which they are formed is shown in Fig. 3.3a, b (see Sect. 3.3). One third of such sSMC were neocentric, the remainder were centric. Again, one third of the patients with an sSMC formed by the McClintock mechanism were clinically normal, as this produced a balanced situation (see Sect. 3.3). Problems appeared in most instances in the cytogenetically unbalanced offspring (e.g., case McCl-08-O-p11.1/2-1; Liehr 2011a).

The McClintock mechanism is thought to be limited to ring-shaped sSMC. However, very similar mechanisms seem to be the basis for the formation of minute-shaped or inverted-duplication-shaped sSMC. In these cases a partial deletion on one of the sSMC sister chromosomes was either compensated for by a minute-shaped sSMC or overcompensated for by an inverted-duplication-shaped one. Examples of neocentric sSMC cases are 01-N-q32/1-1, 09-N-pt12/1-1, 11-N-qt22/1-1, 13-N-qt32.3/2-1, 14-N-qt32.1/1-1, 17-N-qt22/1-1, and 20-N-pt11.2/1-1, and examples of centric sSMC cases are 13-O-q12.2/1-1, 13-U-1, 13-U-2, 18-U-4, 18-U-5, 20-W-p13/4-1, and possibly 0X-CW-6 (Liehr 2011a).

9.2.3   Other Structural Chromosomal Rearrangements

Other kinds of structural chromosomal changes have been observed together with sSMC. They could cause balanced or unbalanced situations.

Rearrangements known as chromosomal polymorphism or heteromorphism, such as pericentric inversion in one chromosome 9, i.e., inv(9)(p11q13) (Starke et al. 2002), were observed in four cases (02-O-p11.2/1-1, 14-O-q11.1/1-19, 14-W-q13.3/1-1, 15-N-qt25/1-4; Liehr 2011a), variant 9ph+ (Starke et al. 2002) was seen in one case (08-O-p11.21/3-1; Liehr 2011a), and a 16qh variant was reported in another case (14-U-17; Liehr 2011a).

Unique inversions in chromosomes other than the sSMC derived from chromosome 15 have also been seen, i.e., inv(2)(q13q36) in case 22-O-q11.1/1-16 and inv (X)(p11.4p22.3) in case 16-O-p11.2/1-1 (Liehr 2011a). Further, there can be balanced translocations independent of sSMC formation, as in cases 15-O-q11.1/1-60 and 15-O-q11.2/1-3 (Liehr 2011a).

Finally, cytogenetically balanced karyotypes involving one of the sister chromosomes of the sSMC could be present (cases 14/22-CO-31, 14-U-1b, 18-CO-7, 21-O-q11.1/1-1, 22-W-q10/1-1, 22-U-3; Liehr 2011a). In the latter cases it was not always clear if the sSMC was connected with the observed rearrangement or not.

If the overall karyotype was unbalanced, the rearrangement additional to the sSMC could obviously be independent of the sSMC formation. That is, the sSMC was derived from another chromosome as in case 01-U-1 having a del(18)(q22) as well as an sSMC(1) or case 19-U-6 with a dup(21)(q22.2q21.1) and an sSMC(19).

Finally, structural chromosomal rearrangements leading to an unbalanced situation could be present involving the same chromosome as the sSMC but with breakpoints other than those present on the sSMC. In those cases a common origin of the sSMC and rearrangement could not be proven, but might be present (cases 06-U-5, 08-U-1, 13/21-U-4, 14/22-U-4, 15-U-3, 15-U-12, 15-U-13, 15-P-q11.2/1-1, 15-P-q12/1-1, 17-W-p13.3/1-1, and 18-U-6 (Liehr 2011a).

9.3   Molecular Aberrations

As mentioned in Sect. 9.2.1, an sSMC need not always be the reason for the clinical problems observed in a patient. Especially when the sSMC leads to a balanced situation and/or is accordingly heterochromatic, one might think of other cryptic genetic aberrations within the patient's genome. In these cases one may think of a complex sSMC (see Chap. 10), a microdeletion syndrome (see Sect. 9.2.1), UPD (see Chap. 6), or if there are special symptoms of disease-causing mutations in special genes.

Six sSMC cases have been reported in which a gene-specific point mutation was disease-causing in the corresponding patients. Fragile X syndrome was found in three instances (cases 04-U-2, 04-U-8, and 22-O-q11.1/1-25; Liehr 2011a). In case 16-U-15 (Liehr 2011a) BWS (OMIM #130650) was diagnosed, caused by a LIT1 methylation defect. Furthermore, an sSMC(7) was accompanied by an adrenogenital syndrome causing homozygote mutation in the CYP21A2 gene (case 07-U-10; Liehr 2011a). Finally, a cobalamin C defect (homozygote c271dupA mutation in the MMACHC gene in 1q34.1) was found in an sSMC(4) carrier (case 04-U-7, Liehr 2011a).

Chapter 10
Complex Small Supernumerary Marker Chromosomes

The formation of complex rearranged sSMC consisting of parts from two or more different chromosomal regions was discussed in Sect. 3.4. Chapter 3 also covered the fact that the most frequently observed complex sSMC are those present in Emanuel syndrome (ES) (see Sect. 5.1), and more than 300 such cases are known. Most often, complex sSMC have a centric minute shape (excluding ES cases there are still more than 60 cases), followed by an inverted duplication shape (15 cases) and ring shape (four cases). Chromosome 15 is most often involved. In most cases two chromosomes are involved, but there are also rare cases of complex rearranged sSMC in which only one chromosome (e.g., 18-W-p11.1/3-1 and 20-O-p11.21/1-1; Liehr 2011a) or even three chromosomes (case 07-U-1; Liehr 2011a) are involved.

A complex sSMC can be suspected if – after (molecular) cytogenetic analysis – a seemingly heterochromatic sSMC is connected with clinical problems. Such sSMC were studied by microdissection and reverse FISH and were found to consist of euchromatin in addition to heterochromatin. In most cases, this euchromatin derived from subtelomeric regions; these subtelomeric regions could be from any of the human chromosomes (Trifonov et al. 2008).

Overall, for complex sSMC fewer than 100 cases are known, excluding ES cases. However, as they are difficult to identify, this group might be bigger than presently known.

T. Liehr, *Small Supernumerary Marker Chromosomes (sSMC)*,
DOI 10.1007/978-3-642-20766-2_10, © Springer-Verlag Berlin Heidelberg 2012

Chapter 11
Small Supernumerary Marker Chromosomes and Tumors

Marker chromosomes, including supernumerary ones, are regularly observed in tumor cytogenetics. However, they are mostly larger than a chromosome 20 of the same metaphase spread, and thus they are not by definition sSMC. In addition, and more importantly, these SMC and sSMC in tumors are normally acquired and nonconstitutional. In other words, there is presently no correlation between sSMC and tumors.

Still, as mentioned in Chap. 7 (see Sects. 7.1, 7.3, 7.9) neocentric sSMC restricted to tumor tissues have been reported. Also, there are a few reports of sSMC restricted to tumor tissues (Isobe et al. 1990; Starke et al. 2001; Garsed et al. 2009). Even though the possibility has been discussed from time to time (Streubel et al. 2001; Wieland et al. 2006; Bakshi et al. 2008), there is no hint that sSMC might by causative for or related to any kind of human cancer (Liehr 2011a).

T. Liehr, *Small Supernumerary Marker Chromosomes (sSMC)*,
DOI 10.1007/978-3-642-20766-2_11, © Springer-Verlag Berlin Heidelberg 2012

Appendix: Patient Organizations in Connection with Small Supernumerary Marker Chromosomes

As underlined by the multiple personal case reports, sSMC patients and parents of sSMC carriers are shocked after receiving the aberrant cytogenetic result. Thus, in this situation they are in need of any kind of reliable specialist support they can get. Besides clinicians being well informed about the sSMC problem, patient support groups are a good source of help. In the following, rare chromosome disorder support groups, in general, and chromosome-specific support groups are listed. We apologize in advance for those not listed, as we did not miss them out them intentionally. Facebook (www.facebook.com) provides a useful source of informal contact and support for families. There is a rapidly growing number of pages on Facebook specific to particular chromosomes or disorders.

A1   Rare Chromosome Disorder Support Groups

A1.1   Unique: Rare Chromosome Disorder Support Group

Website: http://www.rarechromo.org
E-mail: info@rarechromo.org
Helpline: +44-1883-330766
Address: Unique, PO Box 2189, Caterham, Surrey CR3 5GN, UK

Unique offers support and information primarily in English but some information guides on individual rare chromosome disorders are available in languages such as German, French, Spanish, Italian, Danish, Greek, Romanian, Arabic and Dutch. Unique publishes a family-friendly guide to sSMC in English and German.

Having a child with a rare chromosome disorder can be a huge shock and stir up a range of emotions and a desire to learn more about the child's disorder. Almost everyone who works for Unique has been through these experiences. Some parents want to find another, older child with the same disorder as their child. Although this may be possible for some, it does not mean that the two children will develop in the

T. Liehr, *Small Supernumerary Marker Chromosomes (sSMC)*,
DOI 10.1007/978-3-642-20766-2, © Springer-Verlag Berlin Heidelberg 2012

same way. However, just talking to other parents with a child with a rare chromosome disorder can be a great relief and can help to dispel feelings of isolation and "Why me?"

Unique runs a helpline for families and professionals to find out more about specific rare chromosome disorders. It has developed an extensive computerized database detailing the effects of specific rare chromosome disorders among its members. The database can be used to link families on the basis of a specific rare chromosome disorder. Often of more practical benefit, however, is to link families on the basis of problems as they arise, whether these are medical, developmental, behavioral, social, or educational.

Unique also maintains close links with similar groups around the world, thus increasing the range of possible family contacts. Information about a specific rare chromosome disorder can be prepared from the Unique database without revealing the identity of the families concerned. This service is widely used by geneticists and genetic counselors worldwide.

Many local groups and contacts have been formed throughout the UK and in some other countries. Families affected by any rare chromosome disorder can get together locally for support and friendship and to pass on information about local services available. Unique publishes a regular magazine in which families can write about their experiences and exchange information. The group also holds study meetings and conferences in the UK where families and professionals can meet and discuss the latest developments. Unique can also act as a go-between to enable families to participate in any research projects relevant to their child's condition. Whatever a family's specific needs, Unique tries to provide them with tailor-made information and help relevant to their child's disorder.

A1.2   Contact a Family: for Families with Disabled Children

Website: http://www.cafamily.org.uk/rda-uk.html
E-mail: helpline@cafamily.org.uk
Telephone: 0808-8083555 (UK only)
Address: Contact a Family, 209–211 City Road, London EC1V 1JN, UK

Contact a Family provides support, advice, and information for families with disabled children; it is primarily active in the UK.

A1.3   LEONA: Verein für Eltern chromosomal geschädigter Kinder e.V.

Website: http://www.leona-ev.de
E-mail: info@leona-ev.de
Telephone: +49-4421-748669
Address: LEONA e.V. Kreihnbrink 31, 30900 Wedemark, Germany

LEONA is a group of rare chromosome disorder families with 400 members, more than 800 contact families, and representing over 300 syndromes. LEONA is active chiefly in German-speaking countries. Information and help are free of charge.

A1.4   Valentin APAC

Website: http://www.valentin-apac.org
E-mail: contact@valentin-apac.org
Helpline: +33-1-30379097
Address: Valentin APAC, 52, la Butte Eglantine, 95610 Eragny, France

Valentin APAC is the rare chromosome disorder support group for France and the French-speaking countries of Europe. It has more than 3,500 contact families.

A1.5   Unique Danmark

Website: http://www.uniquedanmark.dk
E-mail: formand@uniquedanmark.dk
Telephone: +45-3250-4155
Address: Unique Danmark, c/o Dorte V. Moeller, Blykobbervej 9 st.tv., 2770
   Kastrup, Denmark

Unique Danmark is a rare chromosome disorder support group for families in Denmark.

A1.6   Chromosome Disorder Outreach

Website: http://www.chromodisorder.org
E-mail: info@chromodisorder.org
Telephone: +1-561-3954252
Address: Chromosome Disorder Outreach, Inc., PO Box 724, Boca Raton, FL
   33429–0724, USA

Chromosome Disorder Outreach is a nonprofit organization providing support and information for families caring for a child, teen, or adult diagnosed with a rare chromosome disorder, including chromosome deletions, duplications, rings, inversions, and translocations. Chromosome Disorder Outreach operates primarily in English but has limited material available in Spanish, French, and Italian.

A1.7   Asociaţia Prader Willi din România

Website: http://www.apwromania.ro
E-mail: doricad@yahoo.com
Telephone: +40-260-616585
Address: Asociaţia Prader Willi din România, Str. Simion Bărnuţiu Nr. 97, Bl.
   SB88, ap. 14, Zalău, Sălaj, Romania

Asociaţia Prader Willi din România is a rare-disease-oriented organization specifically created for Romania and provides support in Romanian and English.
Initially it was created by parents of a child with Prader Willi syndrome.

A1.8   Living with Trisomy

Website: http://www.livingwithtrisomy.org/
E-mail: fawna33@mindspring.com
Telephone: not applicable
Address: not applicable

This website in English provides links to family pages offering information and
support to other families living with trisomy. Most of the linked family webpages
will have trisomy support organization links of their own, but this site is specifically
for families to connect with other families through their own webpages.

A1.9   National Organization for Rare Disorders

Website: http://www.rarediseases.org
E-mail: orphan@rarediseases.org
Telephone: 55 Kenosia Avenue, PO Box 1968, Danbury, CT 06813-1968, USA
Address: +1-203-744-0100

The National Organization for Rare Disorders [NORD] is a federation of voluntary
health organizations dedicated to helping people with rare "orphan" conditions.

A2   Chromosome-Specific Support Groups

In the following, mainly different chromosome-specific support groups are listed.

A2.1   Chromosome 7

A2.1.1   Child Growth Foundation

Website: www.childgrowthfoundation.org
E-mail: info@childgrowthfoundation.org
Telephone: +44-20-8995-0257
Address: Child Growth Foundation, 2 Mayfield Avenue, Chiswick, London
   W4 1PW, UK

The Child Growth Foundation aims to support people with growth disorders, including Russell-Silver syndrome, to promote and fund relevant research, to educate both the public and professionals, and to encourage regular monitoring of growth and development criteria.

A2.1.2   Restricted Growth Association

Website: http://restrictedgrowth.co.uk
E-mail: office@restrictedgrowth.co.uk
Telephone: +44 (0)300-111-1970
Address: *see the website*

The Restricted Growth Association supports individuals and families affected by genetic restricted growth conditions, including Russell-Silver syndrome.

A2.1.3   Human Growth Foundation

Website: http://hgfound.org
E-mail: hgf1@hgfound.org
Telephone: +1-800-451-6434
Address: Human Growth Foundation, 997 Glen Cove Avenue, Suite 5, Glen Head,
   NY 11545, USA

The Human Growth Foundation aims to help people with growth and growth hormone disorders, including Russell-Silver syndrome, through research, education, support, and advocacy.

A2.1.4   The MAGIC Foundation

Website: http://www.magicfoundation.org
E-mail: dianne@magicfoundation.org
Telephone: +1-708-383-0808
Address: The MAGIC Foundation, 6645 W. North Avenue, Oak Park, IL 60302, USA

The MAGIC Foundation exists to provide support services for the families of children with a variety of growth conditions and disorders, including Russell-Silver syndrome.

A2.2   Chromosome 9

A2.2.1   9TIPS Trisomy 9 International Parent Support

Website: http://www.trisomy9.org/9tips.htm   or   http://health.groups.yahoo.com/
    group/9tips/
Email: bl737@bellsouth.net or 9tips@yahoogroups.com
Telephone: +1-909-8624470
Address: not applicable

9TIPS is an international support group for families dealing with all variations of duplications (trisomy) of chromosome 9.

A2.2.2   Trisomy 9

Website: http://www.trisomy9.org
E-mail: mandycain@bigpond.com
Telephone: not applicable
Address: not applicable

A2.3   Chromosome 11

A2.3.1   European Chromosome 11 Network

Website: http://www.chromosome11.eu
E-mail: see the website
Telephone: see the website
Address: see the website

The European Chromosome 11 Network is a support group for patients with chromosome 11 disorders, their families, and relatives.

A2.3.2   Beckwith-Wiedemann Syndrome Support Group

Website: http://www.mdjunction.com/beckwith-wiedemann-syndrome
E-mail: contact@mdjunction.com
Telephone: not applicable

Address: not applicable

An online community of patients, family members, and friends dedicated to dealing with BWS together.

A2.3.3   Beckwith-Wiedemann Support Group

Website: http://bws-support.org.uk
E-mail: r.baker881@btinternet.com; rbaker5165@aol.com
Helpline: +44-1258-817573 (evenings); +44-7889-211000 (cell phone)
Address: Beckwith-Wiedemann Support Group, The Drum and Monkey, Hazelbury Bryan, Sturminster Newton, Dorset DT10 2EE, UK

The UK Beckwith-Wiedemann Support Group aims to promote awareness of BWS, to support and encourage research, and to support affected families.

A2.3.4   L'Associazione Italiana Sindrome di Beckwith-Wiedemann

Website: http://www.aibws.org
E-mail: aibws@libero.it
Telephone: +39-345-3121850 (cell phone)
Address: L'Associazione Italiana Sindrome di Beckwith-Wiedemann, Piazza Turati, 3, 21029 Vergiate (VA), Italy

L'Associazione Italiana Sindrome di Beckwith-Wiedemann aims to increase knowledge about BWS and to contribute usefully to the care of patients and their families.

A2.4   Chromosome 12

A2.4.1   PKS Kids

Website: http://www.pkskids.com and http://www.pkskids.ning.com
E-mail: info@pkskids.net
Telephone: not applicable
Address: PO Box 94, Florissant, MO 60302-0094, USA

PKS Kids is a nonprofit organization and is where families and professionals can find the latest about what is happening in the world of PKS. For families and professionals the website is http://www.pkskids.com.

The PKS Kids social and support forum is at http://www.pkskids.ning.com. Families are invited to join the forum, offer help and friendship, and ask others for advice and answers. Sharing photos and videos of children makes it more fun!

A2.4.2 PKS Support Online

Website: http://groups.yahoo.com/group/pks_support
E-mail: not applicable
Telephone: not applicable
Address: not applicable

PKS Support Online is an international support group for sharing information, real-life experiences, and many other aspects of living with or caring for people diagnosed with PKS.

A2.4.2 PK Syndrome Online

Website: http://www.pk-syndrome.org
E-mail: andrea@pk-syndrome.org
Telephone: not applicable
Address: not applicable

This is a PKS lay support group, offering an interface with PKS-related information in English and Italian. It acts as an entry point to reach useful organizations, saving time and reducing frustration.

A2.5 Chromosome 14

A2.5.1 International Association Ring 14

Website: http://www.ring14.org
E-mail: info@ring14.it
Telephone: +39-0522-421037 (office) or +39-340-8681962 (cell phone)
Address: International Association Ring 14, Via Victor Hugo 34, 42123 Reggio Emilia, Italy

A2.6 Chromosome 15

A2.6.1 Dup15q Alliance

Website: http://www.dup15q.org
E-mail: info@dup15q.org
Telephone: +1-855-3871572
Address: Dup15q Alliance, PO Box 674, Fayetteville, NY 13066, USA

Dup15q Alliance, provides family support and promotes awareness, research, and targeted treatments for chromosome 15q duplication syndrome

A2.6.2   Nonsolo15

Website: http://www.idic15.it
E-mail: info@idic15.it
Telephone: +39-0575-583950
Address: Associazione Nonsolo15, Il Castello 15, Papiano, 52017 Stia (AR), Italy

Nonsolo15 operates in Italian and English.

Nonsolo15 is a nonprofit association providing support for families and friends of people affected by isodicentric chromosome 15 syndrome and supporting relevant research. The organization gives advice, information, and support to families with children with isodicentric chromosome 15 syndrome and to their friends; it helps them to meet each other to share experiences and receive support; it maintains an updated website, translating international information about research and therapies and publishing information material; it promotes research and maintains contact with researchers and physicians; it publicizes the experiences of affected families to the Italian community; and it grows every day with families' own children in pain and joy, and helps them to understand and grow too.

A2.6.3   idic15-EU

Website: http://health.groups.yahoo.com/group/idic15eu
E-mail: not applicable
Telephone: not applicable
Address: not applicable

This is a group for European families with a family member with isodicentric chromosome 15 syndrome, using any European language.

A2.6.4   Prader–Willi Syndrome

There are country-specific support groups and associations for PWS in many countries. In the first instance, contact the International Prader-Willi Syndrome Organisation (see below) or look at the links on the New Zealand Association website.

A2.6.5   International Prader-Willi Syndrome Organisation

Website: http://www.ipwso.org
E-mail: ceo@ipwso.org or g.fornaz@alice.it
Telephone: +39-0444-555557

Address: C/-B.I.R.D. Europe Foundation Onlus, via Bartolomeo Bizio, 36023 Costozza (VI), Italy

A2.6.6   Prader Willi Syndrom Vereinigung Deutschland e.V.

Website: http://www.prader-willi.de
E-mail: info@prader-willi.de
Telephone: +49-5141-3747327
Address: Prader Willi Syndrom Vereinigung Deutschland e.V., Am Bruückhorst 2a, 192 29227 Celle, Germany

Prader Willi Syndrom Vereinigung Deutschland e.V. was started by parents of children with PWS and is active chiefly in German-speaking countries.

A2.6.7   Prader-Willi Syndrome Association (UK)

Website: http://pwsa.co.uk
E-mail: admin@pwsa.co.uk
Telephone: +44-1332-365676
Address: Prader-Willi Syndrome Association (UK), 125a London Road, Derby DE1 2QQ, UK

A2.6.8   Prader-Willi Association (USA)

Website: http://www.pwsausa.org
E-mail: national@pwsausa.org
Telephone: 800-9264797 (USA only) or +1-941-3120400
Address: Prader-Willi Association (USA), 8588 Potter Park Drive, Suite 500, Sarasota, FL 34238, USA

A2.6.9   Prader-Willi Syndrome Association (NZ)

Website: http://www.pws.org.nz
E-mail: ceo@pws.org.nz
Telephone: 0800-479743 (toll-free in New Zealand)
Address: Prader-Willi Syndrome Association (NZ), PO Box 258, Silverdale, Auckland 0944, New Zealand

A2.6.10   Angelman Syndrome

There are many country-specific support groups and associations for AS. In the first instance, contact the International Angelman Syndrome Organisation (see below) or look at the New Zealand Association website.

A2.6.11   The International Angelman Syndrome Organisation

Website: http://www.international.angelmansyndrome.org
E-mail: president@angelmansyndrome.org
Telephone: not applicable
Address: not applicable

A2.6.12   ASSERT Angelman Syndrome Support Education and Research Trust

Website: http://www.angelmanuk.org
E-mail: angelmanuk@live.co.uk
Telephone: +44-300-9990102
Address: ASSERT, PO Box 4962, Nuneaton CV11 9FD, UK

A2.6.13   Angelman Syndrome Foundation

Website: http://www.angelman.org
E-mail: info@angelman.org
Telephone: +1-630-9784245
Address: Angelman Syndrome Foundation, 4255 Westbrook Drive, Suite 219, Aurora, IL 60504, USA

A2.6.14   Angelman Syndrome Association of Australia (Inc)

Website: http://www.angelmansyndrome.org
E-mail: president@angelmansyndrome.org
Telephone: not applicable
Address: Angelman Syndrome Association, PO Box 554, Sutherland, NSW 1499, Australia

A2.6.15   Angelman New Zealand

Website: http://www.angelman.co.nz
E-mail: angelman.info@nzord.org.nz
Telephone: +64-7824-7376
Address: Angelman New Zealand, PO Box 128, Ngaruawahia, New Zealand

A2.7   Chromosome 16

A2.7.1   Disorders of Chromosome 16 Foundation

Website: http://www.trisomy16.org
E-mail: doc16foundation@yahoo.com
Telephone: +1-760-6448867
Address: DOC16 Foundation, PO Box 230448, Encinitas, CA 92023-0448, USA

Disorders of Chromosome 16 Foundation is dedicated both to promoting research and providing information on chromosome 16 abnormalities.

The foundation provides information, education, and support for families of children living with a chromosome 16 disorder and to expectant parents confronting a similar diagnosis. The foundation also serves as a resource aiding family, friends, caregivers, and medical professionals in their supportive roles.

A2.8   Chromosome 17

A2.8.1   familyofchromosome17disorders

Website: http://health.groups.yahoo.com/group/familyofchromosome17disorders
E-mail: familyofchromosome17disorders@yahoogroups.com
Telephone: not applicable
Address: not applicable

This is a group started by the parent of a child with a microdeletion syndrome but which is for parents of a child with any type of rare disorder affecting chromosome 17

A2.8.2   Dup-17p11-2, 17p11.2 Duplication

Website: http://health.groups.yahoo.com/group/Dup-17p11-2
E-mail: dup-17p11-2@yahoogroups.com
Telephone: not applicable
Address: not applicable

The 17p11.2 Duplication Yahoo Group is designed to promote communication and the exchange of information about this chromosome disorder among parents, professionals, and researchers.

A2.8.3   Chromo 17 Europe

Website: http://www.chromo17europe.webs.com
E-mail: via the webpage
Telephone: not applicable
Address: not applicable

This is a support group and information site and forum for parents, carers, grandparents, and friends of children with conditions within the 17th chromosome.

A2.9   *Chromosome 18*

A2.9.1   Chromosome 18 Registry and Research Society

Website: http://www.chromosome18.org
E-mail: office@chromosome18.org
Telephone: +1-210-6574968
Address: Chromosome 18 Registry and Research Society, 7155 Oakridge Drive, San Antonio, TX 78229, USA

The Chromosome 18 Registry and Research Society is a lay advocacy organization composed primarily of the parents of individuals with one of the chromosome 18 abnormalities. Its mission is to help individuals with chromosome 18 abnormalities to overcome the obstacles they face so they may lead happy, healthy, and productive lives. The society provides specific information on tetrasomy 18p – see http://www.chromosome18.org/TheConditions/Tetrasomy18p/tabid/129/Default.aspx.

A2.9.2   Chromosome 18 Registry and Research Society (Europe)

Website: http://www.chromosome18eur.org
E-mail: secretary@chromosome18eur.org
Telephone: +44-1236-823455
Address: Bonnie McKerracher, 14 Main Street, Twechar, East Dunbartonshire G65
   9TA, United Kingdom

Chromosome 18 Registry and Research Society (Europe) is a charity set up to
support families whose children are affected by all chromosome 18 disorders. It
provides information and support through its website, and parents who join the
registry have access to a daily listserv, where they can ask questions and receive
advice and encouragement from other families facing the same challenges every
day. The charity aims to hold biannual family conferences. It is closely affiliated
to the Chromosome 18 Registry and Research Society in the USA, the leading
research body into the disorders.

A2.9.3   Tetrasomy18p.ca

Website: http://www.tetrasomy18p.ca
E-mail: tetrasomy18p@yahoo.ca
Telephone: not applicable
Address: not applicable

This is an information-sharing website, set up by a Canadian family.

A2.9.4   Tetrasomy 18p

Website: http://health.groups.yahoo.com/group/tetrasomy18p/
E-mail: tetrasomy18p@yahoogroups.com
Telephone: not applicable
Address: not applicable

This is a private community for those who have loved ones with this rare genetic
syndrome. Tetrasomy 18p is a syndrome based on chromosome 18.

A2.10   Chromosome 20

A2.10.1   Ring Chromosome 20 Foundation

Website: http://www.ring20.org
E-mail: info@ring20.org

Telephone: +44-1708-403620 (UK) or +1-212-8602552 (USA)
Address: Ring Chromosome 20 Foundation, 62 Ravel Gardens, Aveley, Essex
    RM15 4NH, UK or Ring Chromosome 20 Foundation, 1045 Park Avenue,
    New York, NY 10028, USA

A2.11   Chromosome 21

A2.11.1   International Mosaic Down Syndrome Association

Website: http://www.imdsa.org/
E-mail: brandy@imdsa.org
Telephone: +1-513-9886817 (Skype calls from USA only)
Address: International Mosaic Down Syndrome Association, PO Box 354, Trenton,
    OH 45067, USA

International Mosaic Down Syndrome Association is designed to assist any family
or individual whose life has been affected by mosaic Down syndrome, assist in
research, and provide support without regard to race, sex, or religion.

A2.12   Chromosome 22

A2.12.1   Emanuelsyndrome.org or Chromosome 22 Central

Website: http://www.emanuelsyndrome.org or http://www.c22c.org
E-mail: info@emanuelsyndrome.org or steph.stpierre@gmail.com or murney.
    rinholm@c22c.org
Telephone: +1-705-2683099 or +1-919-5678167
Address: Chromosome 22 Central Inc, c/o Stephanie St-Pierre, 338 Spruce Street
    North, Timmins, ON P4N 6N5, Canada or c/o Murney Rinholm, 7108
    Partinwood Drive, Fuquay-Varina, NC 27526, USA

Emanuelsyndrome.org is part of the larger parent group Chromosome 22 Central
(http://www.c22c.org). Chromosome 22 Central provides families with information
and offers opportunities for them to network with others through the Internet and at
family gatherings. It provides information in English and has a contact for inquiries
in Spanish.

A2.12.2   Ring 22

Website: http://health.groups.yahoo.com/group/ring22/
E-mail: ring22@yahoogroups.com
Telephone: not applicable
Address: not applicable

This is a resource for families and friends of people with ring chromosome 22.

A2.13   X and Y Chromosomes

There are many country-specific support groups and associations for TS. In the first instance, log on to the Turner Syndrome Society of the United States website (see below) and follow the links for International Contacts.

A2.13.1   Turner Syndrome Society of the United States

Website: http://www.turnersyndrome.org
E-mail: tssus@turnersyndrome.org
Telephone: 800-3659944 (toll-free in the USA)
Address: Turner Syndrome Society of the United States, 11250 West Road #G,
   Houston, TX 77065, USA

A2.13.2   Turner Syndrome Support Society (UK)

Website: http://www.tss.org.uk/
E-mail: turner.syndrome@tss.org.uk
Helpline: + 44-845-2307520
Address: Turner Syndrome Support Society (UK), 13 Simpson Court, 11 South
   Ave, Clydebank Business Park, Clydebank G81 2NR, UK

The society provides information on TS and the many aspects of living with the condition on a daily basis.

Glossary

This book is about a very complicated topic that can sometimes be hard for the layman to understand. Here an attempt is made to explain some of the most important technical and medical terms related to genes and chromosomes.

Chromosomes and Nomenclature

The human body is made up of billions of cells. Most of the cells contain a complete set of approx. 21,000 genes which act like a set of instructions, controlling growth and development and how the body works. Genes are carried on microscopically small, threadlike structures called chromosomes. There are usually 46 chromosomes, 23 inherited from the mother and 23 inherited from the father. Apart from two sex chromosomes (two X chromosomes for a girl and an X and a Y chromosome for a boy), there are 22 pairs of chromosomes that are numbered from 1 to 22, generally from largest to smallest.

The cytogenetic description of the chromosome set a person carries is expressed by a karyotype. This shorthand code usually states the total number of chromosomes, e.g., 46, followed by the sex chromosomes, e.g., XX or XY, followed by the numbers of cells studied in square brackets, e.g., [15]. A karyotype for a healthy female should normally read as 46,XX[15], and that of a healthy male should read as 46,XY[15]. A girl with Down syndrome describes in most cases the presence of an additional chromosome 21, so the karyotype is written as 47,XX,+21 [15]. If an sSMC is present, the karyotype of a boy reads as 47,XY,+mar[15]. The better an sSMC is characterized, the more complex the karyotype formula becomes. If the sSMC has a ring shape, is derived from chromosome 1, including bands 1p12 to 1q12, and is present in mosaic in a girl, the karyotype would be, e.g., 47,XX,+r(1) (::p12->q12::)[15]/46,XX[35]. The formula looks complicated and can be much more complicated, spanning several lines in a report, if all molecular cytogenetic probes applied are listed. In general, the length of the formula cannot be aligned with the severity of the clinical phenotype to be expected! The advantage of using

T. Liehr, *Small Supernumerary Marker Chromosomes (sSMC)*,
DOI 10.1007/978-3-642-20766-2, © Springer-Verlag Berlin Heidelberg 2012

karyotypes is that they are understood in every country by cytogenetic specialists, irrespective of the language a report is written in. The principles of the nomenclature are summarized in the so-called International System for Human Cytogenetic Nomenclature – ISCN (2009).

Alphabetic List of Terms

- *Acrocentric(s)* are chromosomes without a short p-arm, i.e., chromosomes 13, 14, 15, 21, and 22; as they have similar shapes with the centromere being near one chromosomal end, they are distinguished from all other, so-called nonacrocentric chromosomes.
- *Alleles* are one of two or more forms of a particular gene; alleles differ by their DNA sequences.
- *Alpha fetoprotein* is a major plasma protein produced by the developing baby. Some of this protein passes across the placenta and can be detected in the mother's blood during pregnancy. Testing alpha fetoprotein (AFP) values is a routine test for pregnancy surveillance.
- *Alphoid or satellite DNA* is a repetitive sequence stretch primarily located in the centromeric region of chromosomes. In humans there are specific sequences at almost each centromere.
- *Amniocentesis* is used in prenatal diagnosis of chromosomal abnormalities and fetal infections. Around 10 ml of amniotic fluid containing fetal cells is acquired by needle aspiration under ultrasound control.
- *Anaphase* is one of the five stages of mitosis.
- *Aneusomy* means any numerical deviation from a normal diploid karyotype; it may be a gain or a loss of a chromosome.
- *Array-based comparative genomic hybridization* (aCGH) is a DNA-directed array technique. Array-based methods are gradually replacing microscopy as the preferred approach to chromosome analysis (karyotyping) for children with developmental disorders as they have a higher diagnostic yield.
- *Array techniques* are recently invented methods which allow a high-resolution analysis of human DNA, RNA, and proteins.
- *Ataxia* is a clinical sign implying dysfunction of the parts of the nervous system that coordinate movement, such as the cerebellum. It consists of gross lack of coordination of movement.
- *Atresia* means that a passage in the body is congenitally closed or absent e.g., "gut atresia" means that there is a developmental blockage in the gut.
- *Autosomes* are all human chromosomes from chromosome 1 to chromosome 22 (i.e., all the chromosomes except the X and Y chromosomes, which are known as the sex chromosomes).
- *Azoospermia* describes the fact that a male has no measurable level of sperm in his ejaculate.

- *Centromere* is the narrow part of the chromosome between the short and the long arms. The centromere is the attachment spot for the spindle apparatus during mitosis.
- *Chorion* is a membrane present only during pregnancy between the developing fetus and the mother. Chorionic villi may be biopsied as an alternative to amniocentesis to study the chromosomal makeup of the developing pregnancy in pregnancies undergoing prenatal diagnosis. Chorionic villus sampling is usually undertaken at approximately 11–12 weeks of pregnancy.
- *Chromatids* (two of them) form one chromosome; they are joined at the centromere.
- *Clinodactyly* describes a bend or curvature of a finger.
- *Coarctation* is an abnormal narrowing in a blood vessel.
- *Coloboma* is a gap in one of the structures of the eye, e.g., the iris.
- *Cryptorchidism* is the failure of descent of one testis or both testes into the scrotum.
- *Cyanosis* indictes a lack of oxygen, which may happen during birth or postnatally.
- *Cytoband* is a G bands by trypsin using Giemsa stain (GTG)-light or GTG-dark chromosomal subband.
- *Cytogenetics* is a synonym for chromosome analysis.
- *Cytomegalovirus* belongs to the herpesvirus group.
- *Epicanthus* (adjective: epicanthic) describes a skin fold of the upper eyelid in the inner corner of the eye.
- *Epigenetics* refers to any heritable influence (in the progeny of cells or of individuals) on chromosome or gene function that is not accompanied by a change in DNA sequence, e.g., X-chromosome inactivation, imprinting, centromere inactivation, and position effect variegation.
- *Euchromatin* is genetic material containing actively transcribed/translated genes.
- *Gamete* is a human cell that fuses with another gamete during fertilization, i.e., sperm and oocyte.
- *Gametogenesis* is the process by which gametes form, i.e., spermatogenesis and oogenesis.
- *Genotype* is constituted by the genetic information present in a cell or a person.
- *Gonadoblastoma* is a (benign) tumor derived from different germ cells.
- *Gonosome* is a nonautosomal chromosome, i.e., a sex chromosome; in humans this is X and Y chromosomes.
- *Hemizygosity* means that a chromosomal region in a diploid organism is only present in one copy.
- *Hemorrhage* is the medical term for bleeding.
- *Heterochromatin* is used in this book as genetic material without (active) genes.
- *Heterodisomy* describes a special type of uniparental disomy. In heterodisomy there are two different homologous chromosomes derived from one parent $<->$ isodisomy.
- *Homologous* chromosomes designate a pair of two identical chromosomes.

– *Howell–Jolly* bodies are small DNA-containing inclusions of erythrocytes (red blood cells) and are often present after splenectomy.
– *Hypertelorism* is an abnormally increased distance between two organs – the term is most commonly used to describe wide spacing of the eyes.
– *Hypospadias* is a birth defect in which the opening of the urethra is unusually positioned, e.g., on the shaft of the penis rather than at the tip.
– *Hypothyroidism* is a deficiency of thyroid hormone.
– *Icterus* means jaundice.
– *Intrauterine* means within the uterus.
– *Isochromosome* is a derivative chromosome consisting of two identical short arms or two identical long arms.
– *Isodisomy* describes a special type of uniparental disomy. In isodisomy there are two identical homologous chromosomes derived from one parent $<->$ heterodisomy.
– *Karyograms* are depictions of chromosomes which are sorted by size, centromere position, and banding pattern in a standard format.
– *Karyotypes* describe the number of chromosomes, and what they look like under a light microscope. There is special nomenclature for that description, which ends up in a karyotype formula understandable worldwide if correctly applied.
– *Kinetochore* is the protein structure on chromosomes where the spindle attaches during cell division.
– *Meiosis* (adjective: meiotic) is a special type of cell division necessary for sexual reproduction leading to the formation of gametes.
– *Mesoderm* is one of the three primary germ cell layers in the very early embryo.
– *Microphthalmia* is a developmental disorder of the eye and just means "small eye."
– *Mitosis* is the process by which a eukaryotic cell separates the chromosomes in its nucleus into two identical sets in two nuclei.
– *Monosomy* means that instead of two copies of a chromosomal region only one copy is present.
– *Mosaic/mosaicism* describes the presence of two cell populations with different genotypes in one individual who has developed from a single zygote.
– *Neoplasia* is another word for tumor.
– *Nondisjunction* is the failure of chromosome pairs to separate properly during cell division.
– *Occiput* is the anatomical term for the back portion of the head.
– *Omphalocele* is an abdominal wall defect leading to a smaller or larger bulge containing the intestines at the site of the umbilicus ("tummy button").
– *Oocyte* is the female gamete.
– *Phenotype* is any observable characteristic of an individual.
– *Postzygotic* is the time after the first cell of an individual formed, i.e., the zygote.
– *Scoliosis* means that a person's spine is curved from side to side.
– *Sirenomelia* is a very rare malformation in which the legs are fused together, giving the appearance of a mermaid's tail.
– *Spermatocyte* is a male gamete.

- *Spermatogenesis* is the process during which a male gamete differentiates to a sperm.
- *Stenosis* is an abnormal narrowing in a blood vessel or another tubular structure of the body.
- *Strabismus* is a condition in which the eyes are not properly aligned with each other.
- *Tetrasomy* means that instead of two copies of a chromosomal region four copies are present.
- *Tracheostomy* describes the procedure of making an incision in the trachea to enable a patient to breathe.
- *Trimester* is a time period during pregnancy; the time between fertilization and delivery is divided into three equal parts: the first, second, and third trimesters.
- *Trisomy* means that instead of two copies of a chromosomal region three copies are present.
- *Trophectoderm* is, like the chorion, a membrane present only during pregnancy between the developing fetus and the mother.
- *Zygote* is formed by fertilization of an oocyte by a sperm. It is the single cell from which a multicellular organism is formed.

References

Agrelo R, Wutz A (2010) X inactivation and disease. Semin Cell Dev Biol 21:194–200

Ahn JW, Mann K, Walsh S, Shehab M, Hoang S, Docherty Z, Mohammed S, Mackie Ogilvie C (2010) Validation and implementation of array comparative genomic hybridisation as a first line test in place of postnatal karyotyping for genome imbalance. Mol Cytogenet 3:9

Anderlid BM, Sahlen S, Schoumans J, Holmberg E, Ahsgren I, Mortier G, Speleman F, Blennow E (2001) Detailed characterization of 12 supernumerary ring chromosomes using micro-FISH and search for uniparental disomy. Am J Med Genet A 99:223–233

Arnold J (1879) Beobachtungen über Kernteilungen in Zellen der Geschwülste. Virchows Arch Pathol Anat 78:279–301

Augui S, Filion GJ, Huart S, Nora E, Guggiari M, Maresca M, Stewart AF, Heard E (2007) Sensing X chromosome pairs before X inactivation via a novel X-pairing region of the Xic. Science 318:1632–1636

Babić I, Brajenović-Milić B, Petrović O, Mustać E, Kapović M (2007) Prenatal diagnosis of complete trisomy 19q. Prenat Diagn 27:644–647

Backx L, Van Esch H, Melotte C, Kosyakova N, Starke H, Frijns JP, Liehr T, Vermeesch JR (2007) Array painting using microdissected chromosomes to map chromosomal breakpoints. Cytogenet Genome Res 116:158–166

Bakshi SR, Dave BJ, Sanger W, Brahmbhatt MM, Trivedi PJ, Kakadia PM, Patel SJ (2008) Characterization of a familial small supernumerary marker chromosome in a patient with adult-onset tongue cancer. Cytogenet Genome Res 121:14–17

Baldwin EL, May LF, Justice AN, Martin CL, Ledbetter DH (2008) Mechanisms and consequences of small supernumerary marker chromosomes: from Barbara McClintock to modern genetic-counseling issues. Am J Hum Genet 82:398–410

Bán Z, Nagy B, Papp C, Beke A, Tóth-Pál E, Papp Z (2003) Recurrent trisomy 21 and uniparental disomy 21 in a family. Fetal Diagn Ther 18:454–458

Barber JC (2005) Directly transmitted unbalanced chromosome abnormalities and euchromatic variants. J Med Genet 42:609–629

Barber JC (2011) Chromosome anomaly collection. https://www.som.soton.ac.uk/research/Geneticsdiv/Anomaly%20Register/default.htm. Accessed 10 Jan 2011

Barbi G, Kennerknecht I, Wohr G, Avramopoulos D, Karadima G, Petersen MB (2000) Mirror-symmetric duplicated chromosome 21q with minor proximal deletion, and with neocentromere in a child without the classical Down syndrome phenotype. Am J Med Genet A 91:116–122

Bartels I, Schlueter G, Liehr T, von Eggeling F, Starke H, Glaubitz R, Burfeind P (2003) Supernumerary small marker chromosome (SMC) and uniparental disomy 22 in a child with confined placental mosaicism of trisomy 22: trisomy rescue due to marker chromosome formation. Cytogenet Genome Res 101:103–105

Battaglia A (2008) The inv dup (15) or idic (15) syndrome (tetrasomy 15q). Orphanet J Rare Dis 3:30

Bélien V, Gérard-Blanluet M, Serero S, Le Dû N, Baumann C, Jacquemont ML, Dupont C, Krabchi K, Drunat S, Elbez A, Janaud JC, Benzacken B, Verloes A, Tabet AC, Aboura A (2008) Partial trisomy of chromosome 22 resulting from a supernumerary marker chromosome 22 in a child with features of cat eye syndrome. Am J Med Genet A 146:1871–1874

Beverstock GC, Bezrookove V, Mollevanger P, van de Kamp JJ, Pearson P, Kouwenberg JM, Rosenberg C (2003) Multiple supernumerary ring chromosomes of different origin in a patient: a clinical report and review of the literature. Am J Med Genet A 122:168–173

Bloom SE, Goodpasture C (1975) An improved technique for selective silver staining of nucleolar organizer regions in human chromosomes. Hum Genet 34:199–206

Camacho JPM (2004) B chromosomes in the eukaryote genome. Cytogenet Genome Res 106:147–410

Camacho JPM, Shaw MW, López-León MD, Pardo MC, Cabrero J (1997) Population dynamics of a selfish B chromosome neutralized by the standard genome in the grasshopper *Eyprepocnemis plorans*. Am Nat 149:1030–1050

Carter MT, St Pierre SA, Zackai EH, Emanuel BS, Boycott KM (2009) Phenotypic delineation of Emanuel syndrome (supernumerary derivative 22 syndrome): clinical features of 63 individuals. Am J Med Genet A 149:1712–1721

Caspersson T, Farber S, Foley GE, Kudynowski J, Modest EJ, Simonsson E, Wagh U, Zech L (1968) Chemical differentiation along metaphase chromosomes. Exp Cell Res 49:219–222

Cavani S, Malcarne M, Arlanian A, Stagni L, Piombo G, Baldo C, Scaraglio T, Boggio G, Mogni M, Uras A, Alabiso A, Zucca M, Zerrega G, Dagna Briscarelli F, Pierluigi M (2003) Prenatal and postnatal identification of 93 supernumerary small chromosomes. Ann Genet 46:231 (abstract number 7.19)

Chandley AC, Edmond P, Christie S, Gowans L, Fletcher J, Frackiewicz A, Newton M (1975) Cytogenetics and infertility in man. I. Karyotype and seminal analysis: results of a five-year survey of men attending a subfertility clinic. Ann Hum Genet 39:231–254

Chen CP (2007) Chromosomal abnormalities associated with omphalocele. Taiwan J Obstet Gynecol 46:1–8

Choo KH (1997) Centromere DNA dynamics: latent centromeres and neocentromere formation. Am J Hum Genet 61:1225–1233

Chudoba I, Franke Y, Senger G, Sauerbrei G, Demuth S, Beensen V, Neumann A, Hansmann I, Claussen U (1999) Maternal UPD 20 in a hyperactive child with severe growth retardation. Eur J Hum Genet 7:533–540

Cormier-Daire V, Le Merrer M, Gigarel N, Morichon N, Prieur M, Lyonnet S, Vekemans M, Munnich A (1997) Prezygotic origin of the isochromosome 12p in Pallister-Killian syndrome. Am J Med Genet A 69:166–168

Créau-Goldberg N, Gegonne A, Delabar J, Cochet C, Cabanis MO, Stehelin D, Turleau C, de Grouchy J (1987) Maternal origin of a de novo balanced t(21q21q) identified by ets-2 polymorphism. Hum Genet 76:396–398

Crolla JA (1998) FISH and molecular studies of autosomal supernumerary marker chromosomes excluding those derived from chromosome 15: II. Review of the literature. Am J Med Genet A 75:367–381

Crolla JA, Howard P, Mitchell C, Long FL, Dennis NR (1997) A molecular and FISH approach to determining karyotype and phenotype correlations in six patients with supernumerary marker (22) chromosomes. Am J Med Genet A 72:440–447

Crolla JA, Long F, Rivera H, Dennis NR (1998) FISH and molecular study of autosomal supernumerary marker chromosomes excluding those derived from chromosomes 15 and 22: I. Results of 26 new cases. Am J Med Genet A 75:355–366

Crolla JA, Youings SA, Ennis S, Jacobs PA (2005) Supernumerary marker chromosomes in man: parental origin, mosaicism and maternal age revisited. Eur J Hum Genet 13:154–160

Dalprà L, Giardino D, Finelli P, Corti C, Valtorta C, Guerneri S, Ilardi P, Fortuna R, Coviello D, Nocera G, Amico FP, Martinoli E, Sala E, Villa N, Crosti F, Chiodo F, di Cantogno LV, Savin E, Croci G, Franchi F, Venti E, Donti E, Migliori V, Pettinari A, Bonifacio S, Centrone C, Torricelli F, Rossi S, Simi P, Granata P, Casalone R, Lenzini E, Artifoni L, Pecile V, Barlati S, Bellotti D, Caufin D, Police A, Cavani S, Piombo G, Pierluigi M, Larizza L (2005) Cytogenetic and molecular evaluation of 241 small supernumerary marker chromosomes: cooperative study of 19 Italian laboratories. Genet Med 7:620–625

Daniel A, Malafiej P (2003) A series of supernumerary small ring marker autosomes identified by FISH with chromosome probe arrays and literature review excluding chromosome 15. Am J Med Genet A 117:212–222

Davenport ML (2010) Approach to the patient with Turner syndrome. J Clin Endocrinol Metab 95:1487–1495

Deutsche Gesellschaft für Humangenetik (GfH), Berufsverband Deutscher Humangenetiker e.V. (BVDH) (2007) Genetische Beratung. Medgen 19:452–454

Dewald GW (1983) Isodicentric X chromosomes in humans: origin, segregation behaviour, and replication band patterns. In: Sandberg AA (ed) Cytogenetics of the mammalian X chromosome, part A. Liss, New York, pp 405–426

Dufke A, Walczak C, Liehr T, Starke H, Trifonov V, Rubtsov N, Schoning M, Enders H, Eggermann T (2001) Partial tetrasomy 12pter-12p12.3 in a girl with Pallister-Killian syndrome: extraordinary finding of an analphoid, inverted duplicated marker. Eur J Hum Genet 9:572–576

Dupont JM, Le Tessier D, Baverel F, Rouffet A, Rabineau D (1997) Mosaic isochromosome 20q and normal outcome: a new case ascertained by fluorescence in situ hybridization and a review of the literature. Fetal Diagn Ther 12:283–285

Ellis JR, Marshall R, Penrose LS (1962) An aberrant small acrocentric chromosome. Ann Hum Genet 26:77–83

Engel E (1980) A new genetic concept: uniparental disomy and its potential effect, isodisomy. Am J Med Genet A 6:137–143

Ewers E, Yoda K, Hamid AB, Weise A, Manvelyan M, Liehr T (2010) Centromere activity in dicentric small supernumerary marker chromosomes. Chromosome Res 18:555–562

Fang YY, Eyre HJ, Bohlander SK, Estop A, McPherson E, Trager T, Riess O, Callen DF (1995) Mechanisms of small ring formation suggested by the molecular characterization of two small accessory ring chromosomes derived from chromosome 4. Am J Hum Genet 57:1137–1142

Farfalli VI, Magli MC, Ferraretti AP, Gianaroli L (2007) Role of aneuploidy on embryo implantation. Gynecol Obstet Invest 64:161–165

Felka T, Lemke J, Lemke C, Michel S, Liehr T, Claussen U (2007) DNA degradation during maturation of erythrocytes: molecular cytogenetic characterization of Howell-Jolly bodies. Cytogenet Genome Res 119:2–8

Fickelscher I, Starke H, Schulze E, Ernst G, Kosyakova N, Mkrtchyan H, Macdermont K, Sebire N, Liehr T (2007) A further case with a small supernumerary marker chromosome (sSMC) derived from chromosome 1-evidence for high variability in mosaicism in different tissues of sSMC carriers. Prenat Diagn 27:783–785

Forster T, Roy D, Ghazal P (2003) Experiments using microarray technology: limitations and standard operating procedures. J Endocrinol 178:195–204

Fraga MF, Esteller M (2002) DNA methylation: a profile of methods and applications. Biotechniques 33:632, 634, 636–649

Friedrich U, Nielsen J (1974) Bisatellited extra small metacentric chromosome in newborns. Clin Genet 6:23–31

Froland A, Holst G, Terslev E (1963) Multiple anomalies associated with an extra small autosome. Cytogenetics 2:99–106

Fryer AE, Ashworth M, Hawe J, Pilling D, Pauling M, Maye U (2005) Isochromosome 20p associated with multiple congenital abnormalities. Clin Dysmorphol 14:49–50

Garsed DW, Holloway AJ, Thomas DM (2009) Cancer-associated neochromosomes: a novel mechanism of oncogenesis. Bioessays 31:1191–1200

Genevieve D, Cormier-Daire V, Sanlaville D, Faivre L, Gosset P, Allart L, Picq M, Munnich A, Romana S, de Blois M, Vekemans M (2003) Mild phenotype in a 15-year-old boy with Pallister-Killian syndrome. Am J Med Genet A 116:90–93

Goumy C, Beaufrère AM, Francannet C, Tchirkov A, Laurichesse Delmas H, Geissler F, Lemery D, Dechelotte PJ, Vago P (2005) Prenatal detection of mosaic isochromosome 20q: a fourth report with abnormal phenotype. Prenat Diagn 25:653–655

Graf MD, Christ L, Mascarello JT, Mowrey P, Pettenati M, Stetten G, Storto P, Surti U, Van Dyke DL, Vance GH, Wolff D, Schwartz S (2006) Redefining the risks of prenatally ascertained supernumerary marker chromosomes: a collaborative study. J Med Genet 43:660–664

Haab O (1878) Albrecht Von Graefes Arch Ophthamol 24:257

Hahn M, Dambacher S, Schotta G (2010) Heterochromatin dysregulation in human diseases. J Appl Physiol 109:232–342

Hall T, Samuel M, Brain J (2009) Mosaic trisomy 22 associated with total colonic aganglionosis and malrotation. J Pediatr Surg 44:e9–e11

Höckner M, Pinggera GM, Günther B, Sergi C, Fauth C, Erdel M, Kotzot D (2009) Unravelling the parental origin and mechanism of formation of the 47, XY, i(X)(q10) Klinefelter karyotype variant. Fertil Steril 90:e13–e17

Huang XL, Isabel de Michelena M, Leon E, Maher T, McClure R, Milunsky A (2007) Pallister-Killian syndrome: tetrasomy of 12pter->12p11.22 in a boy with an analphoid, inverted duplicated marker chromosome. Clin Genet 72:434–440

Ilberry PLT, Lee CWG, Winn SM (1961) Incomplete trisomy in a mongoloid child exhibiting minimal stigmata. Med J Austr 48:182–184

Ing PS, Lubinsky MS, Smith SD, Golden E, Sanger WG, Duncan A (1987) Cat-eye syndrome with different marker chromosomes in a mother and daughter. Am J Med Genet A 26:621–628

Iourov IY, Vorsanova SG, Yurov YB (2009) Developmental neural chromosome instability as a possible cause of childhood brain cancers. Med Hypotheses 72:615–616

Isobe M, Sadamori N, Russo G, Shimizu S, Yamamori S, Itoyama T, Yamada Y, Ikeda S, Ichimaru M, Kagan J, Croce CM (1990) Rearrangements in the human T-cell-receptor alpha-chain locus in patients with adult T-cell leukemia carrying translocations involving chromosome 14q11. Cancer Res 50:6171–6175

Kallioniemi A, Kallioniemi OP, Sudar D, Rutovitz D, Gray JW, Waldman F, Pinkel D (1992) Comparative genomic hybridization for molecular cytogenetic analysis of solid tumors. Science 258:818–821

Kalousek DK, Howard-Peebles PN, Olson SB, Barrett IJ, Dorfmann A, Black SH, Schulman JD, Wilson RD (1991) Confirmation of CVS mosaicism in term placentae and high frequency of intrauterine growth retardation association with confined placental mosaicism. Prenat Diagn 11:743–750

Kato T, Inagaki H, Yamada K, Kogo H, Ohye T, Kowa H, Nagaoka K, Taniguchi M, Emanuel BS, Kurahashi H (2006) Genetic variation affects de novo translocation frequency. Science 311:971

Kleefstra T, de Leeuw N, Wolf R, Nillesen WM, Schobers G, Mieloo H, Willemsen M, Perrotta CS, Poddighe PJ, Feenstra I, Draaisma J, van Ravenswaaij-Arts CM (2010) Phenotypic spectrum of 20 novel patients with molecularly defined supernumerary marker chromosomes 15 and a review of the literature. Am J Med Genet A 152:2221–2229

Köhler H, Merkel A, Schmitt U, Zypries B, Scholz O (2009) Gesetz über genetische Untersuchungen beim Menschen (Gendiagnostikgesetz – GenDG) vom 31. Juli 2009. Bundesgesetzblatt I 50:2529–2538

Kosyakova N, Weise A, Mrasek K, Claussen U, Liehr T, Nelle H (2009) The hierarchically organized splitting of chromosomal bands for all human chromosomes. Mol Cytogenet 2:4

Kotzot D (2002) Supernumerary marker chromosomes (SMC) and uniparental disomy (UPD): coincidence or consequence? J Med Genet 39:775–778

Kristoffersson U (2008) Regulatory issues for genetic testing in clinical practice. Mol Biotechnol 40:113–117

Kristoffersson U, Schmidtke J, Cassiman JJ (2010) Quality issues in clinical genetic services. Springer, Berlin

Kumar C, Kleyman SM, Samonte RV, Verma RS (1997) Marker chromosomes in fetal loss. Hum Reprod 12:1321–1324

Kunze J (2009) Wiedemanns Atlas klinischer Sndrome, Phänomenologie – Äthiologie – Differenzialdiagnose, 6th edn. Schattauer, Stuttgart

Kurahashi H, Emanuel BS (2001) Long AT-rich palindromes and the constitutional t(11;22) breakpoint. Hum Mol Genet 10:2605–2617

Lejeune J (1959) Chromosomic diagnosis of mongolism. Ann Genet Sem Hop 1:41–49

Liehr T (2006) Familial small supernumerary marker chromosomes are predominantly inherited via the maternal line. Genet Med 8:459–462

Liehr T (ed) (2009a) Fluorescence in situ hybridization (FISH): application guide. Springer, Berlin

Liehr T (2009b) Small supernumerary marker chromosomes (sSMC): a spotlight on some nomenclature problems. J Histochem Cytochem 57:991–993

Liehr T (2010) Cytogenetic contribution to uniparental disomy (UPD). Mol Cytogenet 3:8

Liehr T (2011a) The sSMC homepage. http://www.med.uni-jena.de/fish/sSMC/00START.htm. Accessed 10 Jan 2011. Also accessible via http://markerchromosomes.wg.am or http://markerchromosomes.ag.vu

Liehr T (2011b) Homepage on multicolor fluorescence in situ hybridization (mFISH) literature. http://www.med.uni-jena.de/fish/mFISH/mFISHlit.htm. Accessed 10 Jan 2011

Liehr T (2011c) Homepage on cases with uniparental disomy (UPD). http://www.med.uni-jena.de/fish/sSMC/00START-UPD.htm. Accessed 10 Jan 2011

Liehr T, Claussen U (2002) Current developments in human molecular cytogenetic techniques. Curr Mol Med 2:283–297

Liehr T, Weise A (2007) Frequency of small supernumerary marker chromosomes in prenatal, newborn, developmentally retarded and infertility diagnostics. Int J Mol Med 19:719–731

Liehr T, Pfeiffer RA, Trautmann U (1992) Typical and partial cat eye syndrome: identification of the marker chromosome by FISH. Clin Genet 42:91–96

Liehr T, Beensen V, Starke H, Hauschild R, Hempell E, Fritsche V, Hoppe C, Grosswendt G, Prechtel M, Ziegler M, Claussen U, von Eggeling F (2001) Tetrasomy 21 due to a de novo Robertsonian translocation t(14;21) and an additional free trisomy 21. Clin Genet 60:83–85

Liehr T, Claussen U, Starke H (2004) Small supernumerary marker chromosomes (sSMC) in humans. Cytogenet Genome Res 107:55–67

Liehr T, Brude E, Gillessen-Kaesbach G, König R, Mrasek K, von Eggeling F, Starke H (2005) Prader-Willi syndrome with a karyotype 47, XY,+min(15)(pter->q11.1:) and maternal UPD 15: case report plus review of similar cases. Eur J Med Genet 48:175–181

Liehr T, Starke H, Heller A, Kosyakova N, Mrasek K, Gross M, Karst C, Steinhaeuser U, Hunstig F, Fickelscher I, Kuechler A, Trifonov V, Romanenko SA, Weise A (2006a) Multicolor fluorescence in situ hybridization (FISH) applied for FISH-banding. Cytogenet Genome Res 114:240–244

Liehr T, Mrasek K, Weise A, Dufke A, Rodríguez L, Martínez Guardia N, Sanchís A, Vermeesch JR, Ramel C, Polityko A, Haas OA, Anderson J, Claussen U, von Eggeling F, Starke H (2006b) Small supernumerary marker chromosomes: progress towards a genotype-phenotype correlation. Cytogenet Genome Res 112:23–34

Liehr T, Starke H, Senger G, Melotte C, Weise A, Vermeesch JR (2006c) Overrepresentation of small supernumerary marker chromosomes (sSMC) from chromosome 6 origin in cases with multiple sSMC. Am J Med Genet A 140:46–51

Liehr T, Utine GE, Trautmann U, Rauch A, Kuechler A, Pietrzak J, Bocian E, Kosyakova N, Mrasek K, Boduroglu K, Weise A, Aktas D (2007a) Neocentric small supernumerary marker chromosomes (ssMC): three more cases and review of the literature. Cytogenet Genome Res 118:31–37

Liehr T, Mrasek K, Hinreiner S, Reich D, Ewers E, Bartels I, Seidel J, Emmanuil N, Petesen M, Polityko A, Dufke A, Iourov I, Trifonov V, Vermeesch J, Weise A (2007b) Small supernumerary marker chromosomes (ssMC) in patients with a 45, X/46, X,+mar karyotype: 17 new cases and a review of the literature. Sex Dev 1:353–362

Liehr T, Mrasek K, Kosyakova N, Ogilvie CM, Vermeesch J, Trifonov V, Rubtsov N (2008a) Small supernumerary marker chromosomes (ssMC) in humans; are there B chromosomes hidden among them. Mol Cytogenet 1:12

Liehr T, Wegner R-D, Stumm M, Joksi G, Polityko A, Kosyakova N, Ewers E, Reich D, Wagner R, Weise A (2008b) Pallister-Killian syndrome. Rare phenotypic features and variable karyotypes. Balk J Med Gen 11:65–67

Liehr T, Ewers E, Kosyakova N, Klaschka V, Rietz F, Wagner R, Weise A (2009) How to handle small supernumerary marker chromosomes in prenatal diagnostics. Expert Rev Mol Diagn 9:317–324

Liehr T, Karamysheva T, Merkas M, Brecevic L, Hamid AB, Ewers E, Mrasek K, Kosyakova N, Weise A (2010a) Somatic mosaicism in cases with small supernumerary marker chromosomes. Curr Genomics 11:432–439

Liehr T, Kosyakova N, Weise A, Ziegler M, Raabe-Meyer G (2010b) First case of a neocentromere formation in an otherwise normal chromosome 7. Cytogenet Genome Res 128:189–191

Liehr T, Ewers E, Hamid AB, Kosyakova N, Voigt M, Weise A, Manvelyan M (2011) Small supernumerary marker chromosomes and uniparental disomy have a story to tell. J Histochem Cytochem (in press)

Los FJ, van Opstal D, van den Berg C, Braat AP, Verhoef S, Wesby-van Swaay E, van den Ouweland AM, Halley DJ (1998) Uniparental disomy with and without confined placental mosaicism: a model for trisomic zygote rescue. Prenat Diagn 18:659–668

Magenis RE, Sheehy RR, Brown MG, McDermid HE, White BN, Zonana J, Weleber R (1988) Parental origin of the extra chromosome in the cat eye syndrome: evidence from heteromorphism and in situ hybridization analysis. Am J Med Genet A 29:9–19

Manvelyan M, Riegel M, Santos M, Fuster C, Pellestor F, Mazaurik ML, Schulze B, Polityko A, Tittelbach H, Reising-Ackermann G, Belitz B, Hehr U, Kelbova C, Volleth M, Gödde E, Anderson J, Küpferling P, Köhler S, Duba HC, Dufke A, Aktas D, Martin T, Schreyer I, Ewers E, Reich D, Mrasek K, Weise A, Liehr T (2008) Thirty-two new cases with small supernumerary marker chromosomes detected in connection with fertility problems: detailed molecular cytogenetic characterization and review of the literature. Int J Mol Med 21:705–714

Mau-Holzmann UA (2005) Somatic chromosomal abnormalities in infertile men and women. Cytogenet Genome Res 111:317–336

Maurer B, Haaf T, Stout K, Reissmann N, Steinlein C, Schmid M (2001) Two supernumerary marker chromosomes, originating from chromosomes 6 and 11, in a child with developmental delay and craniofacial dysmorphism. Cytogenet Cell Genet 93:182–187

McDermid HE, Morrow BE (2002) Genomic disorders on 22q11. Am J Hum Genet 70:1077–1088

McTaggart KE, Budarf ML, Driscoll DA, Emanuel BS, Ferreira P, McDermid HE (1998) Cat eye syndrome chromosome breakpoint clustering: identification of two intervals also associated with 22q11 deletion syndrome breakpoints. Cytogenet Cell Genet 81:222–228

Mears AJ, Duncan AMV, Biegel JA, Budarf ML, Emanuel BS, Siegel-Bartelt J, Greenberg CR, McDermid HE (1994) Molecular characterization of the marker chromosome associated with cat eye syndrome. Am J Hum Genet 55:134–142

Mears AJ, El-Shanti H, Murray JC, McDermid HE, Patil SR (1995) Minute supernumerary ring chromosome 22 associated with cat eye syndrome: further delineation of the critical region. Am J Hum Genet 57:667–673

Medne L, Zackai EH, Emanuel BS (2010) Emanuel syndrome. In: Pagon RA, Bird TC, Dolan CR, Stephens K (eds). GeneReviews. University of Washington, Seattle

Mefford HC, Eichler EE (2009) Duplication hotspots, rare genomic disorders, and common disease. Curr Opin Genet Dev 19:196–204

Michalski K, Rauer M, Williamson N, Perszyk A, Hoo JJ (1993) Identification, counselling, and outcome of two cases of prenatally diagnosed supernumerary small ring chromosomes. Am J Med Genet A 46:88–94

Migeon BR, Jeppesen P, Torchia BS, Fu S, Dunn MA, Axelman J, Schmeckpeper BJ, Fantes J, Zori RT, Driscoll DJ (1996) Lack of X inactivation associated with maternal X isodisomy: evidence for a counting mechanism prior to X inactivation during human embryogenesis. Am J Hum Genet 58:161–170

Morel F, Roux C, Bresson JL (2000) Segregation of sex chromosomes in spermatozoa of 46, XY/47, XXY men by multicolour fluorescence in-situ hybridization. Mol Hum Reprod 6:566–570

Mrasek K, Starke H, Liehr T (2005) Another small supernumerary marker chromosome (sSMC) derived from chromosome 2: towards a genotype/phenotype correlation. J Histochem Cytochem 53:367–370

Mrasek K, Schoder C, Teichmann AC, Behr K, Franze B, Wilhelm K, Blaurock N, Claussen U, Liehr T, Weise A (2010) Global screening and extended nomenclature for 230 aphidicolin-inducible fragile sites, including 61 yet unreported ones. Int J Oncol 36:929–940

Mukherjee AB, Murty VV, Rodriguez E, Reuter VE, Bosl GJ, Chaganti RS (1991) Detection and analysis of origin of i(12p), a diagnostic marker of human male germ cell tumors, by fluorescence in situ hybridization. Genes Chromosom Cancer 3:300–307

Murmann AE, Conrad DF, Mashek H, Curtis CA, Nicolae RI, Ober C, Schwartz S (2009) Inverted duplications on acentric markers: mechanism of formation. Hum Mol Genet 18:2241–2256

Narahara K, Hiramoto K, Murakami M, Miyake S, Tsuji K, Yokoyama Y, Namba H, Ninomiya S, Murakami R, Seino Y (1992) Unique karyotypes in two patients with Prader-Willi syndrome. Am J Med Genet A 42:671–677

Nelle H, Schreyer I, Ewers E, Mrasek K, Kosyakova N, Merkas M, Hamid AB, Weise A, Liehr T (2010) Harmless familial small supernumerary marker chromosome 22 hampers diagnosis of fragile X-syndrome. Mol Med Rep 3:571–574

Nietzel A, Rocchi M, Starke H, Heller A, Fiedler W, Wlodarska I, Loncarevic IF, Beensen V, Claussen U, Liehr T (2001) A new multicolor-FISH approach for the characterization of marker chromosomes: centromere-specific multicolor-FISH (cenM-FISH). Hum Genet 108:199–204

Nietzel A, Albrecht B, Starke H, Heller A, Gillessen-Kaesbach G, Claussen U, Liehr T (2003) Partial hexasomy 15pter–>15q13 including SNRPN and D15S10: first molecular cytogenetically proven case report. J Med Genet 40:e28

Niksic SB, Deretic VI, Pilic GR, Ewers E, Merkas M, Ziegler M, Liehr T (2010) Trisomy 21 with a small supernumerary marker chromosome derived from chromosomes 13/21 and 18. Balk J Med Genet 13:55–58

Ogata T, Kagami M, Ferguson-Smith AC (2008) Molecular mechanisms regulating phenotypic outcome in paternal and maternal uniparental disomy for chromosome 14. Epigenetics 3:181–187

Oliveira RM, Verreschi IT, Lipay MV, Eça LP, Guedes AD, Bianco B (2009) Y chromosome in Turner syndrome: review of the literature. São Paulo Med J 127:373–378

Online Mendelian Inheritance in Man (OMIM) (2011) http://www.ncbi.nlm.nih.gov/omim. Accessed 10 Jan 2011

Oracova E, Musilova P, Kopecna O, Rybar R, Vozdova M, Vesela K, Rubes J (2009) Sperm and embryo analysis in a carrier of supernumerary inv dup(15) marker chromosome. J Androl 30:233–239

Ou Z, Li S, Li Q, Chen X, Liu W, Sun X (2010) Duchenne muscular dystrophy in a female patient with a karyotype of 46, X, i(X)(q10). Tohoku J Exp Med 222:149–153

Pallister PD, Meisner LF, Elejalde BR, Francke U, Herrmann J, Spranger J, Tiddy W, Inhorn SL, Opitz JM (1977) The Pallister mosaic syndrome. Birth Defects Orig Art Ser XIII 3B:103–110

Paoloni-Giacobino A, Morris MA, Dahoun SP (1998) Prenatal supernumerary r(16) chromosome characterized by multiprobe FISH with normal pregnancy outcome. Prenat Diagn 18:751–752

Pardue ML, Gall JG (1970) Chromosomal localization of mouse satellite DNA. Science 168:1356–1358

Parokonny AS, Wang NJ, Driscoll J, Cuccaro M, Wolpert C, Malone BM, Schanen NC (2007) Atypical breakpoints generating mosaic interstitial duplication and triplication of chromosome 15q11-q13. Am J Med Genet A 143:2473–2477

Peltomaki P, Knuutila S, Ritvanen A, Kaitila I, de la Chapelle A (1987) Pallister-Killian syndrome: cytogenetic and molecular studies. Clin Genet 31:399–405

Pierson M, Gilgenkrantz S, Saborio M (1975) Syndrome dit de l'oeil de chat avec nanisme hypophysaire et developpement mental normal. Arch Fr Pediatr 32:835–848

Pietrzak J, Mrasek K, Obersztyn E, Stankiewicz P, Kosyakova N, Weise A, Cheung SW, Cai WW, von Eggeling F, Mazurczak T, Bocian E, Liehr T (2007) Molecular cytogenetic characterization of eight small supernumerary marker chromosomes originating from chromosomes 2, 4, 8, 18, and 21 in three patients. J Appl Genet 48:167–175

Pinkel D, Straume T, Gray JW (1986) Cytogenetic analysis using quantitative, high-sensitivity, fluorescence hybridization. Proc Natl Acad Sci USA 83:2934–2938

Raca G, Schimmenti L, Martin CL (2008) Intrachromosomal duplications of 22q11 are not a common cause of isolated coloboma and coloboma with other limited features of cat eye syndrome. Am J Med Genet A 146:401–404

Ramirez-Duenas ML, Gonzalez GJ (1992) fra(1)(p11), fra(1)(q22) and r(1)(p11q22) in a retarded girl. Ann Genet 35:178–182

Ridler MA, Berg JM, Pendrey MJ, Saldana P, Timothy JA (1970) Familial occurrence of a small, supernumerary metacentric chromosome in phenotypically normal women. J Med Genet 7:148–152

Rivera H, Moller M, Hernandez A, Enriquez-Guerra MA, Arreola R, Cantu JM (1984) Tetrasomy 18p: a distinctive syndrome. Ann Genet 27:187–189

Schachenmann G, Schmid W, Fraccaro M, Mannini A, Tiepolo L, Perona GP, Sartori E (1965) Chromosomes in coloboma and anal atresia. Lancet 288:290

Schinzel A (1991) Tetrasomy 12p (Pallister-Killian syndrome). J Med Genet 28:122–125

Schinzel A (2001) Catalogue of unbalanced chromosome aberrations in man. De Gruyter, Berlin, pp 19–22

Schinzel A, Schmid W, Fraccaro M, Tiepolo L, Zuffardi O, Opitz JM, Lindsten J, Zetterqvist P, Enell H, Baccichetti C, Tenconi R, Pagon RA (1981) The 'cat eye syndrome': decentric small marker chromosome probably derived from a 22 (tetrasomy 22pter;q11) associated with a characteristic phenotype. Report of 11 patients and delineation of the clinical picture. Hum Genet 57:148–158

Schreck RR, Breg WR, Erlanger BF, Miller OJ (1977) Preferential derivation of abnormal human G-group-like chromosomes from chromosome 15. Hum Genet 36:1–12

Schröck E, du Manoir S, Veldman T, Schoell B, Wienberg J, Ferguson-Smith MA, Ning Y, Ledbetter DH, Bar-Am I, Soenksen D, Garini Y, Ried T (1996) Multicolor spectral karyotyping of human chromosomes. Science 273:494–497

Seabright M (1971) A rapid banding technique for human chromosomes. Lancet 2:971–972

Sebold C, Roeder E, Zimmerman M, Soileau B, Heard P, Carter E, Schatz M, White WA, Perry B, Reinker K, O'Donnell L, Lancaster J, Li J, Hasi M, Hill A, Pankratz L, Hale DE, Cody JD (2010) Tetrasomy 18p: report of the molecular and clinical findings of 43 individuals. Am J Med Genet A 152:2164–2172

Sepulveda W, Be C (2008) Partial trisomy 20q in a fetus with hypoplastic nasal bone, mild ventriculomegaly, and short femur. Prenat Diagn 28:868–870

Shaffer BL, Caughey AB, Cotter PD, Norton ME (2004) Variation in the decision to terminate pregnancy in the setting of an abnormal karyotype with uncertain significance. In: Abstract book of the 54th annual meeting of the American Society of Human Genetics, p 494 (abstract number 2756)

Shaffer LG, Slovak ML, Campbell LJ (eds) (2009) ISCN 2009: an international system for human cytogenetic nomenclature. Karger, Basel

Shah K, Sivapalan G, Gibbons N, Tempest H, Griffin DK (2003) The genetic basis of infertility. Reproduction 126:13–25

Shaikh TH, Budarf ML, Celle L, Zackai EH, Emanuel BS (1999) Clustered 11q23 and 22q11 breakpoints and 3:1 meiotic malsegregation in multiple unrelated t(11;22) families. Am J Hum Genet A 65:1595–1607

Sheth F, Ewers E, Kosyakova N, Weise A, Sheth J, Patil S, Ziegler M, Liehr T (2009) A neocentric isochromosome Yp present as additional small supernumerary marker chromosome: evidence against U-type exchange mechanism? Cytogenet Genome Res 125:115–116

Sidwell RU, Pinson MP, Gibbons B, Byatt SA, Svennevik EC, Hastings RJ, Flynn DM (2000) Pure trisomy 20p resulting from isochromosome formation and whole arm translocation. J Med Genet 37:454–458

Smits LJ, de Bie RA, Essed GG, van den Brandt PA (2005) Time to pregnancy and sex of offspring: cohort study. BMJ 331:1437–1438

Soukup S, Neidich K (1990) Prenatal diagnosis of Pallister-Killian syndrome. Am J Med Genet A 35:526–528

Speicher MR, Gwyn Ballard S, Ward DC (1996) Karyotyping human chromosomes by combinatorial multi-fluor FISH. Nat Genet 12:368–375

Stankiewicz P, Brozek I, Helias-Rodzewicz Z, Wierzba J, Pilch J, Bocian E, Balcerska A, Wozniak A, Kardas I, Wirth J, Mazurczak T, Limon J (2001) Clinical and molecular-cytogenetic studies in seven patients with ring chromosome 18. Am J Med Genet A 101:226–239

Stankiewicz P, Kuechler A, Eller CD, Sahoo T, Baldermann C, Lieser U, Hesse M, Gläser C, Hagemann M, Yatsenko SA, Liehr T, Horsthemke B, Claussen U, Marahrens Y, Lupski JR, Hansmann I (2006) Minimal phenotype in a girl with trisomy 15q due to t(X;15)(q22.3;q11.2) translocation. Am J Med Genet A 140:442–452

Stanyon R, Rocchi M, Capozzi O, Roberto R, Misceo D, Ventura M, Cardone MF, Bigoni F, Archidiacono N (2008) Primate chromosome evolution: ancestral karyotypes, marker order and neocentromeres. Chromosome Res 16:17–39

Starke H, Raida M, Trifonov V, Clement JH, Loncarevic IF, Heller A, Bleck C, Nietzel A, Rubtsov N, Claussen U, Liehr T (2001) Molecular cytogenetic characterization of an acquired minute supernumerary marker chromosome as the sole abnormality in a case clinically diagnosed as atypical Philadelphia-negative chronic myelogenous leukaemia. Br J Haematol 113:435–438

Starke H, Seidel J, Henn W, Reichardt S, Volleth M, Stumm M, Behrend C, Sandig KR, Kelbova C, Senger G, Albrecht B, Hansmann I, Heller A, Claussen U, Liehr T (2002) Homologous sequences at human chromosome 9 bands p12 and q13-21.1 are involved in different patterns of pericentric rearrangements. Eur J Hum Genet 10:790–800

Starke H, Mitulla B, Beensen V, Trifonov V, Rubtsov N, Heller A, Ziegler M, Neumann A, Claussen U, Liehr T (2003) First postnatal case of mosaic del(22)/r(22). Prenat Diagn 23:765–767

Stavropoulou C, Mignon C, Delobel B, Moncla A, Depetris D, Croquette MF, Mattei MG (1998) Severe phenotype resulting from an active ring X chromosome in a female with a complex karyotype: characterisation and replication study. J Med Genet 35:932–938

Stefanou EG, Crocker M (2004) A chromosome 21-derived minute marker in a mosaic trisomy 21 background: implications for risk assessments in marker chromosome cases. Am J Med Genet A 127:191–193

Stephane P, Genevieve L (1999) Prenatal supernumerary r(16) chromosome characterized by multiprobe FISH with normal pregnancy outcome. Prenat Diagn 19:181–182

Stochholm K, Juul S, Gravholt CH (2010) Diagnosis and mortality in 47, XYY persons: a registry study. Orphanet J Rare Dis 5:15

Streubel B, Valent P, Lechner K, Fonatsch C (2001) Amplification of the AML1(CBFA2) gene on ring chromosomes in a patient with acute myeloid leukemia and a constitutional ring chromosome 21. Cancer Genet Cytogenet 124:42–46

Sullivan BA, Willard HF (1998) Stable dicentric X chromosomes with two functional centromeres. Nat Genet 20:227–228

Tabor HK, Cho MK (2007) Ethical implications of array comparative genomic hybridization in complex phenotypes: points to consider in research. Genet Med 9:626–631

ten Bosch JR, Grody WW (2008) Keeping up with the next generation: massively parallel sequencing in clinical diagnostics. J Mol Diagn 10:484–492

Teschler-Nicola M, Killian W (1981) Case report 72: mental retardation, unusual facial appearance, abnormal hair. Syndr Identif 7:6–7

Tjio JH, Levan A (1956) The chromosome number of man. Hereditas 42:1–6

Trifonov V, Seidel J, Starke H, Martina P, Beensen V, Ziegler M, Hartmann I, Heller A, Nietzel A, Claussen U, Liehr T (2003) Enlarged chromosome 13 p-arm hiding a cryptic partial trisomy 6p22.2-pter. Prenat Diagn 23:427–430

Trifonov V, Fluri S, Binkert F, Nandini A, Anderson J, Rodriguez L, Gross M, Kosyakova N, Mkrtchyan H, Ewers E, Reich D, Weise A, Liehr T (2008) Complex rearranged small supernumerary marker chromosomes (sSMC), three new cases; evidence for an underestimated entity? Mol Cytogenet 1:6

Tsuchiya KD, Opheim KE, Hannibal MC, Hing AV, Glass IA, Raff ML, Norwood T, Torchia BA (2008) Unexpected structural complexity of supernumerary marker chromosomes characterized by microarray comparative genomic hybridization. Mol Cytogenet 1:7

Tuna M, Knuutila S, Mills GB (2009) Uniparental disomy in cancer. Trends Mol Med 15:120–128

Turleau C (2005) Orphanet cat-eye syndrome, ORPHA195. http://www.orpha.net/consor/cgi-bin/Disease_Search.php?lng=EN&Expert=195&data_id=246&addSigns=1. Accessed 10 Jan 2011

Vermeesch JR, Melotte C, Salden I, Riegel M, Trifnov V, Polityko A, Rumyantseva N, Naumchik I, Starke H, Matthijs G, Schinzel A, Fryns JP, Liehr T (2005) Tetrasomy 12pter-12p13.31 in a girl with partial Pallister-Killian syndrome phenotype. Eur J Med Genet 48:319–327

Vialard F, Molina-Gomes D, Quarello E, Leroy B, Ville Y, Selva J (2009) Partial chromosome deletion: a new trisomy rescue mechanism? Fetal Diagn Ther 25:111–114

Vogel I, Lyngbye T, Nielsen A, Pedersen S, Hertz JM (2009) Pallister-Killian syndrome in a girl with mild developmental delay and mosaicism for hexasomy 12p. Am J Med Genet A 149:510–514

von Beust G, Sauter SM, Liehr T, Burfeind P, Bartels I, Starke H, von Eggeling F, Zoll B (2005) Molecular cytogenetic characterization of a de novo supernumerary ring chromosome 7 resulting in partial trisomy, tetrasomy, and hexasomy in a child with dysmorphic signs, congenital heart defect, and developmental delay. Am J Med Genet A 137:59–64

von Eggeling F, Hoppe C, Bartz U, Starke H, Houge G, Claussen U, Ernst G, Kotzot D, Liehr T (2002) Maternal uniparental disomy 12 in a healthy girl with a 47, XX,+der(12)(:p11–>q11:)/46, XX karyotype. J Med Genet 39:519–521

Voullaire LE, Slater HR, Petrovic V, Choo KH (1993) A functional marker centromere with no detectable alpha-satellite, satellite III, or CENP-B protein: activation of a latent centromere? Am J Hum Genet A 52:1153–1163

Wang NJ, Parokonny AS, Thatcher KN, Driscoll J, Malone BM, Dorrani N, Sigman M, LaSalle JM, Schanen NC (2008) Multiple forms of atypical rearrangements generating supernumerary derivative chromosome 15. BMC Genet 9:2

Warburton D (1984) Outcome of cases of de novo structural rearrangements diagnosed at amniocentesis. Prenat Diagn 4(7):69–80

Warburton D (1991) De novo balanced chromosome rearrangements and extra marker chromosomes identified at prenatal diagnosis: clinical significance and distribution of breakpoints. Am J Hum Genet A 49:995–1013

Warburton PE (2004) Chromosomal dynamics of human neocentromere formation. Chromosome Res 12:617–626

Warburton D, Anyane-Yeboa K, Francke U (1987) Mosaic tetrasomy 12p: four new cases, and confirmation of the chromosomal origin of the supernumerary chromosome in one of the Pallister-Mosaic syndrome cases. Am J Med Genet A 27:275–283

Weber JL (1990) Human DNA polymorphisms and methods of analysis. Curr Opin Biotechnol 1:166–171

Weise A, Mrasek K, Fickelscher I, Claussen U, Cheung SW, Cai WW, Liehr T, Kosyakova N (2008a) Molecular definition of high-resolution multicolor banding probes: first within the human DNA sequence anchored FISH banding probe set. J Histochem Cytochem 56:487–493

Weise A, Gross M, Mrasek K, Mkrtchyan H, Horsthemke B, Jonsrud C, Von Eggeling F, Hinreiner S, Witthuhn V, Claussen U, Liehr T (2008b) Parental-origin-determination fluorescence in situ hybridization distinguishes homologous human chromosomes on a single-cell level. Int J Mol Med 21:189–200

Wieland I, Muschke P, Volleth M, Ropke A, Pelz AF, Stumm M, Wieacker P (2006) High incidence of familial breast cancer segregates with constitutional t(11;22)(q23;q11). Genes Chromosom Cancer 45:945–949

Wolff DJ, Schwartz S (1992) Characterization of Robertsonian translocations by using fluorescence in situ hybridization. Am J Hum Genet A 50:174–181

Wood E, Dowey S, Saul D, Cain C, Rossiter J, Blakemore K, Stetten G (2008) Prenatal diagnosis of mosaic trisomy 8q studied by ultrasound, cytogenetics, and array-CGH. Am J Med Genet A 146:764–769

Wu TF, Chu DS (2008) Epigenetic processes implemented during spermatogenesis distinguish the paternal pronucleus in the embryo. Reprod Biomed Online 16:13–22

Wu YC, Fang JS, Lee KF, Estipona J, Yang ML, Yuan CC (2003a) Prenatal diagnosis of occipital encephalocele, mega-cisterna magna, mesomelic shortening, and clubfeet associated with pure tetrasomy 20p. Prenat Diagn 23:124–127

Wu YC, Yu MT, Chen LC, Chen CL, Yang ML (2003b) Prenatal diagnosis of mosaic tetrasomy 10p associated with megacisterna magna, echogenic focus of left ventricle, umbilical cord cysts and distal arthrogryposis. Am J Med Genet A 117:278–281

Zackai EH, Emanuel BS (1980) Site-specific reciprocal translocation, t(11;22)(q23;q11), in several unrelated families with 3:1 meiotic disjunction. Am J Med Genet A 7:507–521

Zakowski MF, Wright Y, Ricci A Jr (1992) Pericardial agenesis and focal aplasia cutis in tetrasomy 12p (Pallister-Killian syndrome). Am J Med Genet A 42:323–325

Zhang F, Gu W, Hurles ME, Lupski JR (2009) Copy number variation in human health, disease, and evolution. Annu Rev Genomics Hum Genet 10:451–481

Zhou X, Rao NP, Cole SW, Mok SC, Chen Z, Wong DT (2005) Progress in concurrent analysis of loss of heterozygosity and comparative genomic hybridization utilizing high density single nucleotide polymorphism arrays. Cancer Genet Cytogenet 159:53–57

Index

T. Liehr, *Small Supernumerary Marker Chromosomes (sSMC),*
DOI 10.1007/978-3-642-20766-2, © Springer-Verlag Berlin Heidelberg 2012